I0126267

Agricultural Incentives in Sub-Saharan Africa

Policy Challenges

Robert F. Townsend

The World Bank
Washington, D.C.

Copyright © 1999
The International Bank for Reconstruction
and Development/THE WORLD BANK
1818 H Street, N.W.
Washington, D.C. 20433, U.S.A.

All rights reserved
Manufactured in the United States of America
First printing August 1999

Technical Papers are published to communicate the results of the Bank's work to the development community with the least possible delay. The typescript of this paper therefore has not been prepared in accordance with the procedures appropriate to formal printed texts, and the World Bank accepts no responsibility for errors. Some sources cited in this paper may be informal documents that are not readily available.

The findings, interpretations, and conclusions expressed in this paper are entirely those of the author(s) and should not be attributed in any manner to the World Bank, to its affiliated organizations, or to members of its Board of Executive Directors or the countries they represent. The World Bank does not guarantee the accuracy of the data included in this publication and accepts no responsibility for any consequence of their use. The boundaries, colors, denominations, and other information shown on any map in this volume do not imply on the part of the World Bank Group any judgment on the legal status of any territory or the endorsement or acceptance of such boundaries.

The material in this publication is copyrighted. The World Bank encourages dissemination of its work and will normally grant permission promptly.

Permission to photocopy items for internal or personal use, for the internal or personal use of specific clients, or for educational classroom use, is granted by the World Bank, provided that the appropriate fee is paid directly to Copyright Clearance Center, Inc., 222 Rosewood Drive, Danvers, MA 01923, U.S.A., telephone 978-750-8400, fax 978-750-4470. Please contact the Copyright Clearance Center before photocopying items.

For permission to reprint individual articles or chapters, please fax your request with complete information to the Republication Department, Copyright Clearance Center, fax 978-750-4470.

All other queries on rights and licenses should be addressed to the World Bank at the address above or faxed to 202-522-2422.

ISSN: 0253-7494

ISBN: 0-8213-4528-1

Robert F. Townsend is a young professional in the Eastern and Southern African Rural Development Technical Family in the Africa Region of the World Bank.

TABLE OF CONTENTS

PART I: INTRODUCTION

PART II: THE EXTERNAL MARKET ENVIRONMENT

PART III: THE INTERNAL MARKET ENVIRONMENT

PART IV: REMOVING BARRIERS TO IMPROVE AGRICULTURAL INCENTIVES IN SUB-SAHARAN AFRICA

PART V: COUNTRY PROFILES

BOXES

FIGURES

TABLES

FOREWORD

Agricultural growth is the engine of rural development, and for Africa this growth will have to rest on five key pillars; improving the policy regime, technology development and adoption, rural infrastructure, natural resource management and empowerment of rural people via decentralization and participation. The main focus of this study is on the first of these five pillars, which does not mean that the others are not important. Indeed, all get referenced, although not extensively examined, highlighting strong complementarities between them.

The objective is to take stock of current price related policies in several African countries to help develop a stronger consensus on an appropriate policy stance that will improve agricultural incentives and raise average agricultural growth rates and farm incomes. The study serves to update knowledge and to stimulate discussion.

The findings of the study show that during the 1990s, Sub-Saharan Africa has improved it's agricultural policy regime significantly with reasonable exchange rates and more competitive markets in many countries. Where policy regimes have improved, farmers and consumers have benefited. The policy improvements have been accompanied by increasing export production and investment incentives resulting in higher land productivity, increased exports and rejuvenated agricultural growth.

However, despite these improvements there remains a large unfinished policy agenda which needs continual attention from both African countries and the international community. These include: improving certain institutional arrangements for risk management; finding ways to gain improved access to foreign markets; reducing trade barriers; macroeconomic stability; ensuring a favorable institutional environment; controlling new local levies and taxes; improving rural investment; fostering private and public partnerships and encouraging competition and investment in marketing, processing and input supply.

As we move into the 21st century, Africa faces tremendous opportunities for growth, in which agriculture will continue to play a prominent role. Implementing the unfinished policy agenda is critical to realizing these opportunities.

Hans P. Binswanger
Sector Director
Agriculture and the Environment
Africa Region

ABSTRACT

This study examines the state of agricultural incentives in Sub-Saharan Africa, taking stock of the current policy environment and its recent evolution. The study serves to update knowledge and generate discussion to help develop a stronger consensus on the appropriate policies and incentives that will raise agricultural growth. The global price environment is examined together with the macroeconomic, export crop, food crop and fertilizer policies in sixteen African countries. Based on an array of price ratios and policy scores, policy diamonds are constructed as incentive indicators reflecting the state of macroeconomic (monetary, exchange rate and fiscal) and agricultural sector policies (on exports, food crops and fertilizers) relative to a perceived frontier. An attempt is then made to determine the factors inhibiting countries from moving towards this frontier. Both price and non-price factors are identified as constraining factors and several quantitative methodologies, together with a descriptive analysis, were used to isolate their effects. Cross-country differences in the incentive indictors could be explained by differences in macroeconomic and agricultural policy, the extent and quality of the transport infrastructure, the volumes traded on domestic and international markets and the differences in types of crops produced.

The study highlights several continuing policy challenges that Sub-Saharan Africa faces to ensure appropriate agricultural incentives to stimulate growth. These include: coping with agricultural commodity price decline and fluctuation; securing access to foreign markets and in particular meeting the sanitary and phytosantiory requirements; removing continuing domestic trade barriers; stabilizing macroeconomic policies; enhancing the institutional framework and the credibility of rules; removing the remnants of marketing boards in many African countries; removing excessive agricultural taxation and ensuring public rural investment; improving transportation infrastructure; encouraging public and private sector partnerships and dealing with aid in input markets.

ACRONYMS AND ABBREVIATIONS

ADMARC	Agricultural and Marketing Development Corporation (Malawi)
AGRICOM	Agricultural Marketing Enterprise (Mozambique)
AISO	Agricultural Input Supply Corporation (Ethiopia)
AISE	Agricultural Input Supply Enterprise (Ethiopia)
CAISTAB	Caisse de Stabilisation et de Soutien des Prix des Produits Agricoles (Côte d'Ivoire)
CAR	Central African Republic
CCZ	Cotton Company of Zimbabwe
CEE	Central and Eastern European
CFA franc	Currency known in West Africa as the 'franc de la Communauté financière d'Afrique' and known in Central Africa as the 'franc de la Coopération financière en Afrique centrale'
CFDT	Compagnie Française de Développement des Fibres Textile
CIS	Commonwealth of Independent States
CIF	Costs, Insurance, Freight
CIDT	Compagnie Ivoirienne de Développement des Textiles (Côte d'Ivoire)
CMBL	Coffee Marketing Board Limited (Uganda)
CMC	Cocoa Marketing Company (Ghana)
CMDT	Compagnie Malienne de Développement des Textiles (Mali)
COCOBOD	Cocoa Board (Ghana)
CPSP	Caisse de Péréquation et de Stabilisation des Prix (Senegal)
CSA	Central and South America
DAP	Diammonium phosphate
DIMA	Agriculture Inputs and Equipment Department (Ministry of Agriculture, Burkina Faso)
DRDR	Direction Régionale du Développement Rural (Togo)
EAP	East Asia and Pacific
FAO	Food and Agriculture Organization
f.o.b.	Free on Board
GMB	Grain Marketing Board (Zimbabwe)
GCU	Gambia Cooperative Union
ICA	International Commodity Agreement
KPCU	Kenya Planters Co-operative
LAC	Latin America and Caribbean
ME	Middle East
MENA	Middle East and North Africa
NCB	Nigeria Cocoa Board
NMS	New Marketing System (Mozambique)
NGO	Non-Government Organization
NCPB	National Cereals and Produce Board (Kenya)
NTB	Non-Tariff Barrier
JICA	Japan International Cooperation Agency
OECD	Organisation for Economic Co-operation and Development
ONASA	Office National d'Appui à la Sécurité Alimentaire (Benin)

ONCPB	Office National de Commercialisation des Produit de Base (Cameroon)
ONCC	Office National du Cafe et du Cacao (Cameroon)
PBC	Produce Buying Company (Ghana)
SAED	Société d'Aménagement et d'Exploitation des Terres du Delta (Senegal)
SAFEX	South African Futures Exchange
SENCHIM	Marketing Company of Industries Chimiques du Sénégal (ICS)
SFFRFM	Smallholder Farmers Fertilizer Revolving Fund of Malawi
SOCAPALM	Société Camerounaise de Palmeraies (Cameroon)
SODECOTON	Société de Développement du Coton (Cameroon)
SODEFITEX	Cotton Development Agency, Senegal (*Société pour le Développement des Fibres Textiles)*
SOFITEX	Société des Fibres et des Textiles (Burkina Faso)
SONAPRA	Société Nationale pour la Promotion Agricole (Benin)
SONACOS	Société Nationale de Commercialisation des Oléogineux (Senegal)
SONAGRAINES	SONACOS subsidiary responsible for seed distribution and groundnut collection (Senegal).
SOTOCO	Société Togolaise du Coton (Togo)
SSA	Sub-Saharan Africa
TAMA	Tobacco Association of Malawi
TCMB	Tanzania Coffee Marketing Board
TFC	Tanzania Fertilizer Company
TSP	Triple superphosphate
UCDA	Uganda Coffee Development Authority
UMEOA	West African Economic Union
WTO	World Trade Organization
ZFC	Zimbabwe Fertilizer Company

ACKNOWLEDGMENTS

The report was prepared by Robert Townsend under the guidance of Charles Humphreys and Hans Binswanger. The work also benefited significantly from comments and inputs from the steering committee which included Gershon Feder, Alberto Valdes, John McIntire, Albert Nyberg and Graeme Donovan. Inputs into the country profiles were gratefully received from Vicente Ferrer-Andreu, Steve Jaffee, Jorge Munoz, Tekola Dejene, John McIntire, Patrick Labaste, Ismael Ouedraogo and Nicaise Ehoue. Three external reviewers, Bruce Gardner (University of Maryland), Carl Eicher (Michigan State University) and Ademola Oyejide (University of Ibadan) provided valuable comments and critiques on earlier drafts. The study acknowledges the Japanese Trust Funds for providing resources to employ Tatshusi Adachi as a summer intern who made valuable contributions to the fertilizer section of the study. Brigitte Aflalo provided efficient editorial support to the final drafts.

The report was written in the Eastern and Southern Africa Rural Development Department (AFTR1) and benefited from extensive comments and suggestions from members of that department, especially from the Sector Manager, Sushma Ganguly.

When this report was prepared Robert Townsend was a consultant to the Eastern and Southern Africa Rural Development Family of the World Bank while on leave of absence from the University of Pretoria. The views in the report are those of the author and do not necessarily reflect the views of the World Bank or the University of Pretoria. The author alone is responsible for any errors and omissions.

EXECUTIVE SUMMARY

The importance of generating agricultural growth in Africa is well known. The sector accounts for about 35 percent of the continents GDP, 40 percent of its exports and about 70 percent of employment (World Bank, 1997). Poverty continues to be widespread in the rural areas with 59 percent of the rural population living below the poverty line (Ali and Thorbecke, 1997). Food needs of the continent are also expected to double in the next 20-30 years.

Generating agricultural output and productivity growth remain key components for improving economic and rural development in Sub-Saharan Africa.

Agricultural Growth In Sub-Saharan Africa Has Recently Shown Some Encouraging Trends.

In recent years, agricultural growth in SSA has shown some encouraging signs. Between 1990 and 1997, twenty five countries had real agricultural GDP growth rates over 2 percent, with 12 of these countries having growth rates of over 4 percent. Over the last five years (1993-97), an additional 6 countries joined this group. These recent trends are a significant improvement from the 1980s when only 3 countries had annual growth rates exceeding 4 percent.

Agricultural growth in Sub-Saharan Africa has improved in the 1990s...

Agricultural productivity growth has been less impressive. Most of the growth in cereal production has been from an expansion in area planted which increased at 3.4 percent per annum between 1980 and 1997. Over the same period, cereal yields increased in 24 countries with 7 of these having annual growth rates of more than 2 percent. Measures of multi-factor productivity show similar growth to yields. Between 1980 and 1996, only 4 countries experienced a multi-factor productivity growth of more than 2 percent.

...but is still lagging other developing regions.

Explaining Agricultural Output and Productivity Growth: Many explanations have been developed to account for differences in agricultural growth among countries and over time. Explanations of the relatively poor growth in Africa have typically included poor resource endowments - poor land quality, endemic livestock diseases, human diseases, landlockedness, civil strife and political unrest - and poor policies. Gender inequality has more recently been highlighted as having significant negative effects. More formal theoretical explanations have been separated into a number of functional headings which include elementary, static, dynamic and agglomerate engines of growth (Fafchamps, 1998).

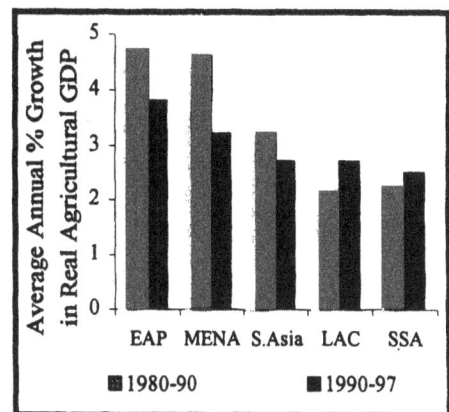

Source: World Development Indicators, 1999.

Markets are central to these growth engines. Firstly, they provide a way to allocate resources, ensuring the highest value of production and maximum consumer satisfaction. Secondly, they can stimulate growth by promoting technological innovation and increased supply and demand (Scarborough and Kydd, 1992).

Past market interventions in Sub-Saharan Africa distorted agricultural price incentives resulting in an inefficient allocation of resources and comparative advantage was not realised. Removing these distortions can provide a significant source of growth from both short term gains (a one time stimulus to growth) and longer term gains (from increased private sector investment). Dynamic or long term growth is more typically induced by the accumulation of productive resources (human and physical) and by technological change resulting in an outward shift of the production possibilities frontier, as opposed to moving towards an existing one.

Changes In Efficiency As Well As Technical Progress Are Important For Productivity Growth.

Agricultural policies that distort incentives, distort allocative efficiency and hence agricultural productivity. Measures of agricultural productivity change have generally assumed that gains in productivity are largely the result of technological progress. Recent methodologies allow us to examine this hypothesis by decomposing productivity measures into changes in efficiency and a technological progress component (Fried *et al*, 1993). Thus productivity growth can be defined as the net change in output due to the change in efficiency and technical change, where the former is understood to be the change in how far an observation is from the frontier of technology and the latter is understood to be shifts in the production frontier (Grosskopf, 1993).

The decomposition is important from a policy perspective as a slowdown in productivity growth due to increased inefficiency suggests different policies than a slowdown due to a lack of technical change. An example is that slow productivity growth due to inefficiency may be due to institutional barriers constraining the diffusion of innovation. In this case, policies to remove these barriers may be more effective in improving productivity than policies directed at innovation. While several African countries appear to be on the efficiency frontier, a

Continued improvements in efficiency can be a significant source of growth for Sub-Saharan Africa...

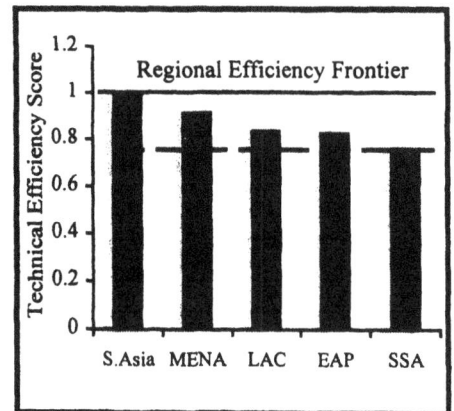

Source: See Box A1 for calculation methodology.

comparison with other developing regions suggests that Africa is lagging behind. These results suggest that significant agricultural productivity gains can be realised by improving efficiency, an improvement which can be induced by policy change.

This study takes stock of current policies, as well as their recent evolution, to update knowledge and to generate discussion to help develop a stronger consensus on an appropriate policy stance that will improve agricultural incentives and raise average agricultural growth rates and farm incomes. Several price (agricultural, trade and macroeconomic policy) and non-price (transportation infrastructure) factors are examined. Following this analysis, several policy challenges to improve these incentives in Africa are highlighted.

A Stock-Take Of Current Policies In Sixteen African Countries Suggests Many Have Significant Space To Improve Agricultural Incentives And Efficiency.

Over the past two decades, policy reform in SSA has been widespread with significant impacts on agricultural incentives. An attempt was made in the study to determine the extent of individual country progress by developing policy diamonds.

The construction of the (pricing) diamond is based on a blend of price ratios and qualitative assessments with each apex representing an indicator of macroeconomic, export crop, food crop and fertilizer policy respectively. The macroeconomic policy scores are developed from an aggregate of monetary, fiscal and exchange rate policies, as in the Adjustment in Africa study (World Bank, 1994). The export crop policy indicator is the ratio of the producer price to the border price. The food crop policy is the ratio of the producer price to the world price, supplemented with qualitative assessments, and the fertilizer policy score is an aggregate score of trade and pricing policies. Two diamonds are constructed, an outer diamond, which is taken to represent the perceived policy frontier (along the apex) and an inner diamond, taken to be the current policy stance of each individual country. In many countries, the distance between these two diamonds is significant, suggesting room for further improvements in agricultural incentives. Indeed, one of the questions the study addresses is what are the constraining economic factors that are preventing countries (inner diamond) from moving towards the frontier (outer diamond) ?

...and removing policies that distort agricultural incentives could improve efficiency and investment to facilitate this growth.

While these four indicators only represent a partial measure of the overall agricultural policy stance (credit, labor, etc., are not included), they nevertheless provide some useful measures of four key incentive policy areas (macroeconomic, export crop, food crop and fertilizer). To supplement this static picture, a more dynamic overview is provided by examining the recent evolution of these policies, highlighting the changes in the external and internal environment in which agriculture operates. Indeed, there has been much progress in improving agricultural incentives in Africa.

The Improvements In Price Incentives During The 1990s Were Due To More Favorable World Prices, Continuing Improvements In Macroeconomic Policies And, In Some Countries, Due To An Increase In The Producer Share Of The Border Price.

The decomposition of agricultural prices into these three elements provides useful insights into the dominant causes of changing price incentives. The trends in each of these components are examined.

Favorable World Prices In The 1990s Provided A Welcomed Reprieve From The Long-Term Downward Trends.

The international prices of the major agricultural exports from Africa have been declining over the past several decades. The importance of these declining trends on African economies is exacerbated by agriculture's dominance which accounts for about 40 percent of exports. Even within agriculture, several primary crops dominate. This high export concentration exposes many African economies to severe terms of trade fluctuations from commodity price shocks.

Favorable real world commodity prices since 1990 have reversed the declining external terms of trade for agriculture in Sub-Saharan Africa.

The commodity boom from 1990 to 1996/97 resulted in a 20 percent increase in the real aggregate world price of Sub-Saharan Africa's agricultural exports. Real world grain crop prices increased by 14 percent while fertilizer prices increased by 8 percent. These favorable price trends improved agriculture's external terms of trade position in many African countries. The external barter terms of trade for agriculture in SSA, measured as the ratio of the price of agricultural exports to the price of food imports and to the manufacturing unit value, improved at a rate of 1.6 and 0.9 percent per annum respectively between 1990 and 1996. The corresponding external net income

Despite these recent favorable prices...

...their long-term downward trend seems to be firmly in place.

terms of trade improved at an annual (growth) rate of 4.2 and 3.7 percent respectively.

This upward movement in world prices was temporary with a continuation of the downward secular trend in the late 1990s. The long-term world price trends towards lower real commodity prices looks firmly set in place. These trends suggest that Africa must produce agricultural commodities at lower cost (adopt new technologies) or its position in world markets will continue to erode.

African countries will have to produce agricultural commodities at lower cost or their position on world markets will continue to erode.

Macroeconomic Policies Have Continued To Improve.

Distorted macroeconomic policies have had a significant impact on African agriculture and in some cases these effects have exceeded the effects of sectoral policy (Schiff and Valdes, 1992). Faced with falling export earnings, worsening balance of payments, mounting debts and declining economic growth, many countries in Sub-Saharan Africa adopted a wide range of policy reforms. Initial reforms sought to correct for macroeconomic imbalances. Significant improvements were made during the 1980s but fiscal balances were still fragile, inflation was above international levels and the parallel market premium for foreign exchange remained high. Using the same criteria as the Adjustment in Africa study (World Bank, 1994), the macroeconomic policy stance of African countries was assessed.

Fiscal policies have shown a significant improvement since 1990-91. The CFA countries appear to have made greater improvements than the non-CFA countries. A contributing factor to this may be the 50 percent devaluation of the CFA franc in 1994 which provided higher domestic currency export revenues.

Macroeconomic policies have continued to improve in the 1990s...

Monetary policy in Sub-Saharan Africa has focused primarily on maintaining low rates of inflation and adequate levels of real interest rates. Countries appear to have been more successful in the past at achieving these monetary policy objectives than their fiscal policy goals. There was a limited improvement in monetary policy between 1990 and 1996/1997. This may be due to the inflationary pressure from significant currency devaluations.

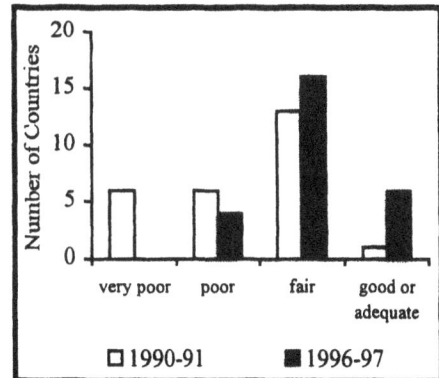

Source: World Bank (1994) and World Bank and IMF data.

Exchange rate policy has improved substantially between 1990-91 and 1996-97 with most of African countries reducing their parallel market exchange rate premiums. The fiscal, monetary and exchange rate policy stance were combined to create a measure of the *overall macroeconomic policy* environment which has improved significantly between 1990/91 and 1996/97. Some countries, however, continued to have poor macroeconomic policies which has led to high levels of macroeconomic instability. The reliability of the institutional framework (credibility of rules) also defines the macroeconomic environment and in Africa, this is lower than in most of the other developing regions (Brunetti, *et al* 1998).

Exchange Rate Pass-Through To Producer Prices Has Improved In Countries With Open Trade Policies.

Between 1990 and 1997, the upward trend in real world commodity prices and improved macroeconomic policies resulted in significant increases in real producer prices of African export crops. However, the currency devaluations have only been fully passed through to farmers as higher prices in countries with open trade policies. The pass-through has been inhibited in countries with highly regulated and controlled markets.

...but currency devaluations have only been fully transmitted to farmers in countries with open trade policies.

The low level of pass-through in many African countries suggests that market inefficiencies exist which have prevented farmers from receiving the full benefits of both favorable world price trends and macroeconomic adjustments. As already highlighted, these inefficiencies become particularly important when analyzing agricultural growth in African countries.

The Current State Of The Producer's Share Of The World Price For Export Crops, Food Crops And Fertilizer Can Be Explained By A Number Of Price And Non-Price Factors.

Export Crop Policies, Markets and Prices.

Producer's Share of the Border Price: The distribution of the producer's share of the f.o.b. price in Sub-Saharan Africa shows a large variation. Most of the farmers receive shares between 40 and 70 percent of the f.o.b. price. This contrast indicates large cross-country differences in the cost of moving the product from the farm gate to the port (taxes, transportation, marketing margins, etc.). In many countries, these high costs are created by policy distortions.

There are large cross-country differences in the producer's share of the border price for export crops...

...which can largely be explained by differences in agricultural policies, macroeconomic policies, rural infrastructure, volumes of crops traded and crop type.

Econometric evidence in this study suggests that the cross-country differences in the producer's share of the f.o.b. price can largely be explained by differences in agricultural and macroeconomic policies, the density and quality of rural roads, volumes of crops traded and crop type.

Market Interventions: Controlled marketing systems continue to distort market price signals in many countries. Four existing marketing systems can be identified in Sub-Saharan Africa, i) the free market systems; ii) the *Caisse de Stabilisation* (e.g. cocoa in Côte d'Ivoire until 1998/99); iii) Marketing Boards (e.g. cocoa in Ghana) and iv) Parastatals (e.g. cotton in West Africa). Farmers under the latter three receive a low share of the border price with interventions in physical handling, price setting, taxation and marketing costs and margins. Exchange rate pass-through to producer prices have subsequently been inhibited in these two systems. The free market system has resulted in substantially higher prices for farmers and lower fiscal costs in the absence of financing the marketing boards or stabilization funds.

Price Stabilization: Many of the current market interventions focus on stabilizing prices, and indeed this argument has been used to justify the existence of many marketing boards and stabilization funds. While prices have been more stable with these interventions, the risk benefits of the more stable prices do not appear to have outweighed the costs of the lower prices offered under the stabilization programs.

Macroeconomic Policies: These have a direct effect on producer prices through exchange rates and money supply and an indirect effect through real interest rates. High real interest rates were found to have a significant negative impact on the producer share of the border price. This result could be explained by the higher cost of borrowing for the private trader who passed-on these higher costs to farmers as lower prices. However, it must be stressed that simple access to credit (which relates to the depth and breadth of the banking system, the rule governing recovery and bankruptcy etc) has also been highlighted as and inhibiting factor.

Infrastructure: Evidence suggests that it is not only the lack of roads that decreases the producer price-border price ratio, but also the quality of these roads. Feeder roads in rural areas remain scarce and are in poor condition in most Sub-Saharan African countries. Transportation is a particular problem for landlocked countries with large differences between coastal prices and border prices. High road tolls in neighboring countries have raised transports costs significantly in many landlocked

State controlled agricultural marketing systems continue to distort price signals in many countries.

While prices have been more stable with interventions, the risk benefits of the more stable prices do not appear to have outweighed the costs of the lower prices offered under the stabilization programs.

countries. Unofficial road blocks erected by local authorities to collect money have also been common.

Volumes Supplied: Development of efficient markets requires volume and consistency in supply and demand. Indeed, private sector entry into storage, transportation and marketing of agricultural products requires some assurance of supply to induce investment. Where high volumes are traded, farmers receive a higher share of the border price and economies of scale reduce transaction costs.

Crop Type: Coffee seems to be a special case, with the producer's share of the border price of this crop being consistently lower than for other African export crops. These prices appear to be consistently lower across all producing countries.

Food Crop Policies, Markets And Prices.

Food markets in Sub-Saharan Africa have been extensively deregulated, induced by the rising costs incurred by state marketing boards. This financial burden was worsened by the operational inefficiencies inherent in these state systems. Unreliable input deliveries and crop payment, the existence of parallel markets and instability of purchases and sales also played a role. An increasing pessimism about the motives and results of state intervention in agricultural markets had developed while there was an optimism in the ability of the private sector to efficiently organize these markets (Lipton, 1991; Jones, 1994).

Liberalization has improved food market integration in many countries...

...but many markets in SSA remain isolated and large transaction costs erode incentives.

Private Trader Emergence: In some African countries, the private sector has been slow to respond to market liberalization. This investment response by private traders is largely determined by market location, area covered by traders, liberalization of marketing policies, favorable exchange rate policies (devaluation increased incentives to invest in the domestic marketing activities at the expense of importers), the profitability of trading activity and the education level and experience of the traders (Badiane *et al*, 1997).

Grain market reforms have induced private trader investments...

Price Response: Removal of policies which set grain, particularly maize, prices in Africa at levels well above world prices has resulted in real producer price declines in many countries, particularly in Southern Africa (e.g. South Africa, Malawi and Zimbabwe). Consumer prices of grain and grain meal have also declined (e.g. Mali, Ghana, Ethiopia and Kenya) with the better transmission of declining real world food prices to domestic markets with the removal of trade restrictions. Particularly in Eastern and Southern Africa, the opening up of markets to private traders has mitigated some of the adverse effects of the declining subsidies. Lower grain processing costs have reduced the wedge between producer and consumer prices (Jayne and Jones, 1997).

Mobility barriers continue to inhibit widespread private sector entry into food markets. These barriers vary at different stages of the marketing chain. *Capital access* can be a serious barrier to entry into subgroups of the marketing chain requiring substantial start-up costs or inventories. The result is relatively closed niches (ie: long-haul transport and interseasonal storage). *Political risk* is also a barrier for private entrepreneur entry into activities with large political uncertainty. *Access to spare parts and energy supplier networks* also provide a significant barrier. The lack of availability of electricity in many rural areas prevents development of rural enterprises such as grain mills. These mills are usually positioned near large municipalities where these services are available. In many countries the natural economies of scale in long-haul motorized transport are accentuated by poor road conditions and limited spare parts and fuel availability. Road blocks have also added to these mobility constraints. (Barrett, 1997).

...but mobility barriers continue to inhibit widespread private sector entry into all marketing niches.

Transportation: High transportation costs have been the source of high food prices in many African countries. In some countries, investment in transportation and storage has been inhibited by the continuation of policies such as pan-territorial pricing structures. Unofficial road blocks erected by local authorities to collect money are also common (e.g. Niger and Madagascar). Road and transportation infrastructure tends to be better and more developed in the West African countries with lower transportation costs. Removing these barriers to entry into these marketing niches may improve market integration and efficiency, thus reducing food prices. There is some evidence that market liberalization has reduced price instability and improved market integration in some countries (e.g. Benin, Ghana, Ethiopia and Malawi) [Dercon, 1995, Badiane *et al* 1997].

Transport costs remain particularly high in some countries and continue to inhibit incentives.

Trade Regimes: In several African countries licenses for grain trade are required which has impeded private imports and exports, reducing their potential to stabilize food supplies and prices (e.g. Burkina Faso, Kenya and Nigeria).

Fertilizer Policies, Prices And Markets.

Fertilizer Prices: Real farm-gate fertilizer prices have increased in most of the countries analyzed. In several countries these prices have increased by more than 100 percent. The increase in the real world fertilizer price, domestic currency devaluation and the removal of subsidies have been the main contributors to this effect.

Currency devaluations have not been consistently passed on to farmers as corresponding higher prices in many countries.

Exchange Rate Pass-Through: In contrast to export crops, if fertilizer prices have not increased as much as the currency devaluation, then farmers benefit. This lack of pass-through could be the result of increased efficiency in the domestic market. Indeed, in several African countries, currency devaluations do not appear to have resulted in proportional price increases (e.g. Zimbabwe, Kenya and Côte d'Ivoire).

Econometric evidence in this study suggests that the cross-country differences in the farm-gate fertilizer price relative to the world price can largely be explained by differences in the distance to international fertilizer markets, trade and macroeconomic policies and the volume of fertilizer traded.

Cross-country variation in the fertilizer farm-gate to world price ratio can largely be explained by differences in...

Transportation and Trade Volumes: Sub-Saharan African countries face high internal and external transport costs exacerbated by the small fertilizer volumes most of these countries import. These small trading shares also weaken negotiating power. Transport costs can sometimes constitute 50 percent of the operating costs of traders. The poor quality of roads further raises fertilizer costs. Local fertilizer traders tend to buy and sell fertilizer within the same season or within a few months with limited investment in storage. Volumes and consistency of demand play a large role in their decisions.

...distances to international fertilizer markets, transportation infrastructure, trade policy macroeconomic policy and volumes of fertilizer traded.

Trade Restrictions and Regulations: Tariffs remain high in several countries which are reflected in higher domestic fertilizer prices. Non-tariff barriers and trade controls restrict fertilizer market entry and competition, contributing to high margins and prices in many countries. Several countries have a restrictive list of fertilizers that can be imported which denies farmers the opportunity of using nutrient-wise, more cost effective fertilizers (e.g. Benin and Ghana). Fertilizer regulations have also been restrictive. Farmers have been cut off from fertilizer markets in neighboring countries, forcing importers to consider countries as a separate and small market (Gisselquist, 1998). Private traders in many countries are not able to bring in fertilizers without prior government approval involving time consuming and costly procedures to satisfy multiple agencies.

Emergence of Private Traders: In most countries, many multi-national companies entered the fertilizer markets after liberalization. Despite this entry, parastatals continue to operate and dominate the fertilizer industry in many countries.

Fertilizer trading differs from grain trading in that the former is characterized by higher levels of working capital, more bulkiness in purchases, and more involvement of multi-national institutions and large companies. Higher skills may be required for this activity. The large financial requirements and the seasonal nature of the market remains a limiting factor to competition and a barrier to the expansion of small-holder companies and new entrants (Badiane *et al*, 1997).

Fertilizer Aid: Sub-Saharan Africa receives extensive fertilizer aid which has had an inhibiting effect on the development of sustainable domestic markets. In many countries, fertilizer aid programs are usually characterized by a lack of competition in procurement and distribution resulting in higher costs of fertilizers. Delivery is slow and the system is vulnerable to mismatching of inputs with needs. Countries reliant on fertilizer aid are also exposed to supply uncertainties.

Fertilizer aid to Africa has caused significant market distortions and in many countries all fertilizer imported is in the form of aid.

Fertilizer Use: Increasing fertilizer use in Africa would require improving not only the profitability of fertilizer use but also the quantity and quality of human resources, financial liquidity, market access, household assets and extension services.

Removing Barriers To Improve Agricultural Incentives.

The study has shown that the external environment and policy regimes have improved in many African countries. However, agricultural incentives in a significant number of countries continue to be inhibited. Transportation costs remain high, traded volumes remain low, private agents have not entered sufficiently into input and output market and the degree of market development and competition remains low. Tariff and non-tariff barriers to agricultural trade continue to be high inside Africa, and in global markets, with improving access to OECD

African countries continue to face significant policy challenges to improve agricultural incentives; these relate to:

markets being a major challenge. Thus, evidence suggests that there continues to be significant room to improve agricultural incentives. This improvement will require focused attention on the macroeconomic environment, trade and market access and on domestic agricultural and rural policies.

The Macroeconomic Environment.

...the macroeconomic environment

Stabilizing Macroeconomic Policy: While macroeconomic polices have improved significantly in African countries, macroeconomic instability has recently been high in many countries (Nigeria, Ghana and Zimbabwe). Subsequent improvements of poor macroeconomic policies (e.g. Ghana and Nigeria) remains fragile and uncertain. This instability and fragility has an adverse effect on investor confidence and growth.

Enhancing the Institutional Framework and the Credibility of Rules: Evidence suggests that the credibility of reforms in Sub-Saharan Africa is poor with unpredictable changes in rules and policies (Brunetti *et al*, 1998). Improving credibility or the reliability of the institutional framework within a country is a critical requirement to encourage private sector activity and investment.

Trade And Market Access.

...trade and market access

Improving Access to Foreign Markets: African countries need to play a greater role in future international trade negotiations to put its concerns on the table and prevent international agreements which unduly disadvantage Africa. To enhance bargaining power in these negotiations, one approach may be to empower the regions (SADC and UEMOA) to put forward Africa's case. Africa will also have to meet the international sanitary and phytosanitary standards, which will become more important to gaining market access.

Removing Domestic Trade Barriers: African countries also need to re-examine their internal trade policies. Import tariffs are higher in Africa than in any other region of the world. Large import tariffs on transportation equipment, machinery and agricultural materials impose higher costs on exporters who use these imports as intermediate inputs in production. This cost disadvantage erodes competitiveness and inhibits growth. Non-tariff barriers in fertilizer markets continue to be widespread which inhibits private sector entry and competition in these markets. High average import duties create a general anti-export bias for the economy discouraging gains in efficiency and diversification.

Coping with World Commodity Price Decline and Fluctuations: The real world agricultural commodity price decline seems firmly in place with high price volatility (booms and busts). Useful strategies to cope with the cyclical trends include adopting prudent monetary and fiscal policies, avoiding higher export taxes, avoiding cartels, using hedging instruments and prudent financial management in times of commodity price booms (Varangis *et al*, 1995). Useful strategies to cope with secular trends include export diversification and more rapid adoption of new technologies.

Fostering Regional Integration: Sufficient volumes and consistency in supply and demand are a requirement for widespread private sector development. In Africa, these volumes, particularly for fertilizer and grain trade, are typically small. Regional integration may provide an avenue to ensure these conditions of supply and demand. This pooling of the rather small markets will enabling more bulk production and purchase of raw materials and facilitate the realization of economies of scale.

Domestic Price And Non-Price Factors.

...the domestic policy environment

Removing Remnants of Marketing Boards: Market interventions such as marketing boards and the *caisse* system are maintained in many African countries. These interventions continue to constrain producer prices by inhibiting the pass-through of currency devaluations to producer prices. These inhibiting effects need to be removed to allow markets to become more competitive.

Taxation and Public Rural Investment: High levels of agricultural taxation needs to be reduced and taxation instruments, such as marketing boards, have generated little government revenue and have led to high deadweight losses. It is, however, obvious that agriculture, as a major sector in many African economies, will have to continue to contribute to government revenues. The key principles that need to be applied to future taxation are nondiscrimination, minimization of negative efficiency impacts, effectiveness of fiscal capture and capacity to implement (Binswanger *at al*, 1999). Consumption taxes may be considered as an alternative revenue-generating mechanism, where administrative capacity is adequate. Where export taxes are justified as substitutes for income tax, the excessively high rates in many countries need to be reduced.

Even though African agriculture was highly taxed in the past, public investment in agriculture and rural development was limited. These investment levels and allocations need to be significantly improved for the much needed capital accumulation in rural areas. One way to improve the effectiveness of public investments would be through fiscal decentralization. Africa has a record of having the one of the most highly centralized political, fiscal and institutional systems for rural development in the world (McLean *et al*, 1999). While these investments are important they must be supported with continual governments spending (i.e. a recurrent budget) to maintain the investments (e.g. road maintenance). Africa does not appear to have done particular well on maintaining its rural investments.

Improving Transport Infrastructure: High transportation costs are one of the key contributors to low producer crop prices and high agricultural inputs prices. Alleviating these transport costs could be achieved directly through investment in roads or indirectly through the reduction of local monopolies, removal of road tolls and lowering of taxes on spare parts.

Fostering Public and Private Sector Partnerships: Greater collaboration and understanding needs to be developed between the public and private sector with clearly defined roles for each. Governments can improve private sector activity by providing a range of public goods. These include investments in infrastructure (roads and electricity), ensuring rule of law and defending open entry and competition. Also in collaboration with the private sector, Governments should make sure market information is as widely disseminated as possible.

Areas For Further Work.

The complexities and inter-linkages within agriculture make it extremely difficult to cover all aspects of agricultural incentives in one study. While providing some insight into the macroeconomic environment, trade and market access and several domestic price and non-price factors, these are by no means the only issues. Further work on agricultural incentives should examine issues of agricultural labor and gender inequality (Blackden and Bhanu, 1999), rural finance, tenure security, agricultural technology and public rural investment.

PART I

INTRODUCTION

1.   AGRICULTURAL GROWTH

Since the early recognition that agriculture can play a central role in generating overall economic development, much attention has been given to identifying ways to accelerate its growth. The common notion in the economic literature is that agriculture plays a pivotal role in the early stages of structural transformation. This perception has induced economists to systemize the process of economic growth into a framework of sequential stages (Rostow, 1990) with agricultural growth viewed as essential to induce the industrialization process (Mellor, 1995). Indeed, there have been very few low-income countries that have achieved rapid nonagricultural growth without corresponding rapid growth in agriculture. Agriculture's role in economic development was highlighted, most noticeably by Johnston and Mellor (1961), to include agriculture as a source of food supply and raw materials, a supplier of foreign exchange, a supplier of labor for industrial employment, a market for non-agricultural output and a source of surplus for investment. Within these roles, one often underplayed contribution is the strength of the backward and forward linkages[1] agriculture has with itself and other sectors of the economy. These linkages generally become stronger with development (Vogel, 1994) and play a key role in agricultural led industrialization (Adelman, 1984).

In Sub-Saharan Africa, agriculture is central to economic growth and the reduction of poverty. The sector accounts for about 35 percent of the continents GDP, 40 percent of its exports and about 70 percent of employment (World Bank, 1997). Recent evidence also suggests that growth stemming from the linkage effects, resulting from increases in African farm incomes, has been underestimated by many previous studies (Delgado et al, 1998). Poverty continues to be widespread in the rural areas with 59 percent of the rural population living below the poverty line[2] (Ali and Thorbecke, 1997), thus stimulating agricultural growth will be a key to poverty reduction strategies in Africa. Food needs are also expected to double in the next 20-30 years.

TRENDS IN AGRICULTURAL OUTPUT AND PRODUCTIVITY GROWTH

An historical view of overall economic growth is provided by Maddison (1996) who presents striking evidence of Africa's poor performance. His cross-regional study suggests that Africa grew at 0.6 percent per annum between 1820 and 1992, which was half the rate of world growth. Five time periods were examined and the strongest growth in Africa was achieved between 1950 and 1970, a period of late colonial rule and early independence. In the post independence era, overall economic growth seems to have slowed and as agriculture is the dominant sector in most African countries, this slow growth is a reflection of agricultural performance. Indeed, over the last 40 years, the fortunes of African agriculture have varied significantly (Figure 1.1). During the 1960s, agriculture grew at about 3 percent per annum, which exceeded population growth at the time. In the following decade, growth declined substantially to about one percent per annum. During this era, the variability of growth across African countries also declined suggesting that the slowdown was widespread across the continent. The

---

[1] As an example – as agricultural production uses inputs from other sectors, such as machinery and fertilizer, an expansion of the agricultural sector should result in an expansion of the industries supplying these inputs to agriculture (*backward linkage*). As sectors use agricultural output as an input, such as cereals for processed foods, the increase in supply of cereals may induce an expansion in the production of processed foods that use cereal as an input (*forward linkage*).

[2] $26 per month per person using 1993 survey data.

performance of agriculture seems to have improved over the recent decades, with growth accelerating over the 1980s and 1990s.

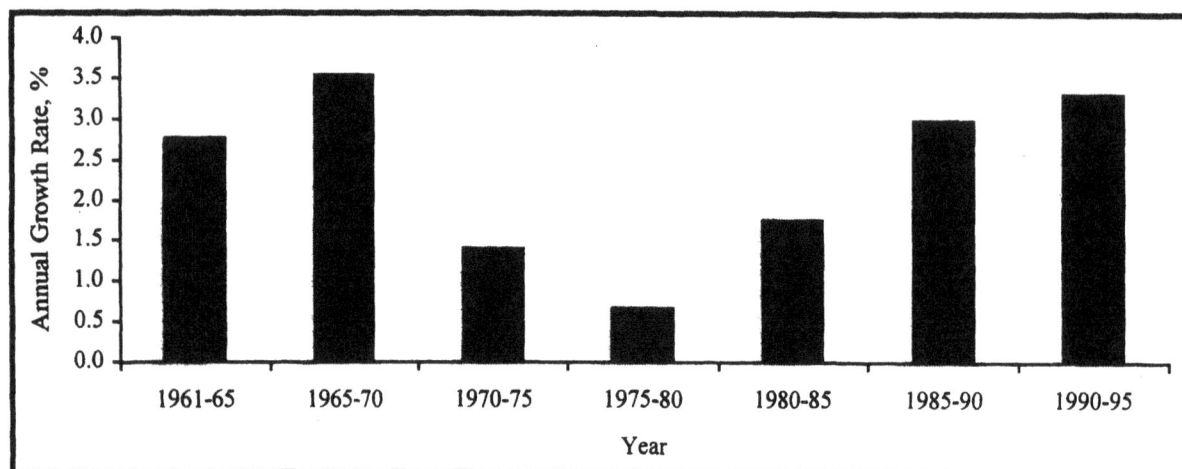

Figure 1.1: Trends in Sub-Saharan African Agricultural Growth, 1961-1995.
Source: FAO data.

The diversity of geography across Sub-Saharan Africa has induced some researchers to group countries into several geographic regions, namely West, Sahelian, Central, East and Southern Africa (FAO, 1986; Bromely, 1995, Block, 1995, See Table A1 in the appendix for the countries included in the regional grouping). These cluster groups have experienced significant differences in growth (Table 1.1). Agriculture in West, Sahelian and Southern Africa grew slowly during the 1970s while accelerating through the 1980s and 1990s. Central Africa experienced strong growth in the 1970s followed by a slow-down in the 1980s and renewed growth in the 1990s. East Africa's growth was relatively slow in the 1970s but doubled during the 1980s, an increase which was largely maintained through the 1990s.

Table 1.1: Regional Differences in Agricultural Growth.

Growth Indicators	Year	West Africa	Sahelian Africa	Central Africa	East Africa	Southern Africa	Sub-Saharan Africa
Real Annual	1970-80	0.6	0.4	2.4	1.0	1.3	1.1
Agricultural GDP	1980-90	2.3	2.6	2.2	2.2	2.1	2.3
Growth (%)	1990-97	2.7	3.1	3.1	2.0	2.5	2.5

Source: World Bank data.

Using similar country groupings, Block (1995) examines trends in total factor productivity for agriculture, finding the strongest growth for West Africa during the 1980s. The emerging pattern from Table 1.1 seems to suggest a hint of regional growth convergence over the last 40 years with similar regional growth rates in the 1990s. When comparing these regions one must bear in mind that even within a region there are large country variations with at least one country in each region (excluding East Africa) having an agricultural growth rate exceeding 4 percent per annum between 1990 and 1997 (Table A1 in the appendix).

Encouraging signs are evident from the improving growth in the in the 1990s. Between 1990 and 1997, 25 countries had real agricultural GDP growth rates over 2 percent, with 12 of these countries

4

having growth rates of over 4 percent (Benin, Cameroon, Chad, Guinea, Guinea-Bissau, Equatorial Guinea, Lesotho, Malawi, Mauritania, Mozambique, Namibia and Togo) [3]. Over the last five years (1993-97) six more countries joined this group (Angola, Côte d'Ivoire, Ethiopia, Mali, South Africa, Zimbabwe). These recent trends are a significant improvement from the 1980s when only three countries had annual growth rates exceeding 4 percent (Benin, Guinea-Bissau and Togo). (Appendix Table A1).

Productivity growth has been less impressive. Between 1980 and 1997, cereal yields increased in 24 countries and seven of these had growth rates of more than 2 percent (Benin, Ghana, Guinea, Guinea Bissau, Mauritania, Cameroon, Central African Republic and Mauritius). Most growth in cereal production was due to area expansion, which increased at 3.4 percent per annum for the same period. Seventeen out of 31 countries examined, show that agricultural value added per worker increased from 1979-81 to 1992-94. Measures of multi-factor productivity show a growth pattern similar to that of yields. Between 1980 and 1996, only four countries of the thirty five examined experienced a multi-factor productivity growth of more than 2 percent (Ghana, Nigeria, Gabon, Uganda) (Appendix 1, Box A1, Table A1).

While aggregate agricultural growth in Sub-Saharan Africa has improved in the 1990s, it is still lagging behind other developing regions. While having the lowest growth rates of the five regions depicted in Figure 1.2, Latin America and the Caribbean (LAC) and Sub-Saharan Africa (SSA) are the only two regions to have exceeded their 1980 growth rates in the 1990s. Explaining differences in growth among countries, geographical regions and continents has been the challenge of many studies of economic growth. A brief review of some of the explanations, and more specifically of those provided for Africa, will be highlighted in the next section.

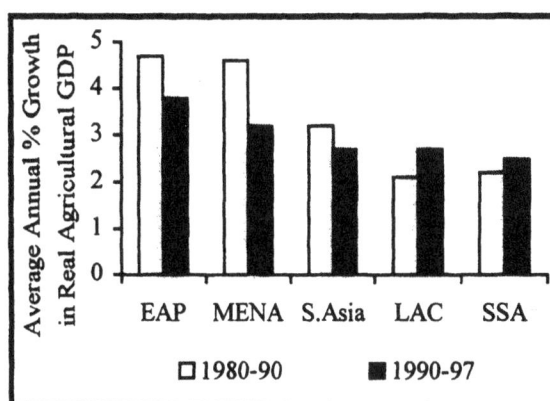

Figure 1.2: Comparisons of Agricultural Growth by Region.
Source: World Development Indicators, 1998.

EXPLANATIONS OF AGRICULTURAL GROWTH

Many models have been formulated to quantify the alternative sources of agricultural growth. Typically, early efforts included only the traditional (land and labour) and modern (fertilizer and machinery) inputs as in the Hayami and Ruttan (1971) formulation. More recently, this list of explanatory

---

[3] Some of these growth rates may be overestimated due to the extensive drought in 1992 which had a significant impact on output, especially in Southern Africa. Agricultural output was thus increasing from a low base.

variables has been expanded to include general and technical education, agricultural research and infrastructure (transport and communications).

Apart from these explanations, Africa's adverse resource endowments have been highlighted as a major constraint to both agricultural and overall growth. Many studies comparing growth rates between Africa and Asia highlight this as one of the primary inhibiting factors for Africa. These include poor land quality (Voortman *et al*, 1998, Donovan and Casey, 1998), endemic livestock diseases (Coetzer *et al*, 1994), human diseases (Bloom and Sachs, 1998), landlockedness (Bloom and Sachs, 1998; Collier and Gunning, 1997) and low population density (Hayami and Platteau, 1997).

Voortman *et al* (1998), highlight the realities of Africa's land ecology. Their study focuses on comparisons with Asia, noting that in Africa there is a prevalence of soils derived from a particular geological formation (metamorphic Basement Complex rock) which is different from the alluvial and volcanic soils (younger and richer in micro-nutrients) prevalent in Asia. They suggest that this difference is one of the primary causes for the lack of replication of the Green Revolution in Africa[4]. Sub-Saharan Africa also has a considerably wider spectrum of infectious animal diseases than any other region of the world (Coetzer *et al*, 1994). Earlier outbreaks of epizootic diseases[5] on the continent were contained, and in some cases eradicated, but their prevalence and distribution in recent years has increased (Thomson, 1997)[6]. In the past, Africa has been extraordinarily vulnerable to many infectious human diseases[7] (malaria, yellow fever, river blindness to name a few) which has been attributed to the natural environment prevalent on the continent (Bloom and Sachs, 1998).

Africa also has an inordinate number of landlocked countries. This situation results in high external transport costs which limits the profitability of many agricultural activities. Bloom and Sachs (1998) highlight the poor access to coasts, also stressing the limited number of navigable rivers in Africa. Their cross-country regressions show these factors to be significant variables influencing growth. Hayami and Platteau (1997) also argue that Africa's poor resource endowment, in particular the abundance of land and low population density, have delayed agricultural growth by increasing transportation costs, suppressing competitive products, factors and credit markets, inhibiting technology adoption and raising the costs of agricultural and social services. Their argument is that the adverse

---

[4] Green Revolution technology requires using high yielding varieties with reliable water supply, in combination with macro-nutrient fertilizer applications to which these varieties are responsive (Nitrogen, Phosphorous and Potassium).

[5] The impact of livestock diseases can typically been separated into two broad groups: erosive diseases and epizootic or transboundary diseases. Of these two groups, the epizootic or transboundary diseases are by far the most important in terms of having the ability to *threaten* large numbers of livestock and thereby the livelihoods of people over wide geographic areas. An outbreak of these diseases result in huge losses. The erosive diseases have less severe results but are a *continuous* problem and the direct losses to the farmer can be substantial.

[6] A recent example of the impact of an epizootic disease is in Botswana. In 1995, an outbreak of cattle lung disease (Contagious Bovine Pleuropneumonia) occurred in the Northern part of the country. The spread of the disease throughout Botswana would have resulted in an appreciable loss of beef exports due to the non-compliance with international food safety regulations. A decision was taken to eradicate the disease through a slaughter policy as a result of which over 300,000 cattle were killed. This event had a tremendous impact on the livelihoods of rural communities in the infected area and can largely explain the negative growth of Botswana's agriculture between 1990-97 (Table A1 in the appendix).

[7] The spread of HIV/AIDS in Africa has been particularly severe and projections suggest that the disease will cause life expectancy to return to 1950 levels wiping out 40 years of development.

6

resource endowments bear much of the responsibility for the failures of agriculture and rural development.

Many agricultural growth studies use these factors to explain the poor economic performance of Africa. Models of agricultural production have included land development costs (Haley and Abbott, 1986), life expectancy and road density (Craig, Pardey and Roseboom, 1994), demographic pressure, expansion of arable land (Frisvold and Lomax, 1991), irrigation, rainfall, and disasters (Ghura and Just, 1992) and infrastructure (Thirtle *at al*, 1995), as explanatory variables. More recently, civil strife and political unrest have also been included and have been shown to be a significant explanatory variable of the poor overall growth performance of Africa (Easterly and Levine, 1995). Gender inequality in education and employment has also been shown to reduce per capita growth in Sub-Saharan Africa (Blackden and Bhanu, 1998).

In a study of overall growth, Sachs and Warner (1998), like Hayami and Platteau (1997), find that resource endowments are a significant variable in explaining cross-country variations of economic growth. However, their findings also suggest that poor economic policies have played an especially important role in causing the poor growth performance in Africa, the most important of these being Africa's lack of openness to international markets[8]. Despite the climatic, geographic and demographic trends, all of which were estimated to hinder growth, Sachs and Warner (1998) estimate that Africa could have achieved per capita growth of 4.3 percent per annum between 1965 and 1990, as opposes to 0.8 percent, if fast-growth policies had been adopted, particularly if it had improved its openness to international markets.

In the context of agriculture, policy has long been highlighted as a key component to enhancing (inhibiting) growth (World Bank 1981), and, since the early 1980s, it has been widely used as an explanatory variable. Several studies have examined the effects of macro policy reforms on agricultural incentives (Jaeger and Humphreys, 1988), on the output of export crops (Jaeger, 1992) and on food crop production (Lamb and Donovan, 1992)[9]. The increased focus on policies was a consequence of the economic crisis and the indebtedness of African countries. Many adopted structural adjustment programs that included policy measures such as elimination of overvalued currencies, price adjustments in the direction of import and export parity, reduction in industrial protection and direct government intervention in the market through a process of liberalization, a decrease of the expenditure burden of governments and redressing fiscal policy (World Bank, 1994). The improved price incentives which were deemed to result from these programs, primarily for export crops, were expected to induce supply and improve efficiency (allocative efficiency and comparative advantage). There has subsequently been a plethora of literature evaluating these adjustment policies.

Some of the literature has focused on the adverse price incentives and excessive government interventions as being the primary constraints to growth (ie: the main explanatory variables for poor performance). Conversely, this emphasis on 'getting prices right' has been criticized as excessive (price fundamentalism) [Lipton, 1987; Krishna, 1990] with arguments that lack of adequate technology and

---

[8] Sachs and Warner (1998) deem an economy to be open to trade if it satisfies five tests (1) average tariff rates below 40%; (2) average quota and licensing coverage of imports of less than 40%; (3) a black market exchange rate premium of less than 20%; (4) no extreme controls (taxes, quotas and state monopolies) on exports and (5) not considered a socialist country by the standard in Kornai (1992).

[9] This is by no means and exhaustive list.

not prices are the most binding elements (Mellor, 1984). Others have stressed the inadequate institutional, human, capital and physical infrastructural environment. Indeed Binswanger and Pingali (1989), in their analysis of agricultural performance in Sub-Saharan Africa find all of these to be important. They attribute agriculture's poor performance to 'faulty pricing policies, inadequate infrastructure and poor institutional development" (p. 48), as well as the wrong technology priorities. While it is recognized that both agricultural price policy and technology are critical to increasing agricultural growth, the emphasis placed on each differs. The price and non-price factor debate was eloquently summarized by Krishna (1990) who called for a balanced view of the role of these factors in promoting growth.

This balance has indeed been stressed extensively in the literature (Eicher, 1989, Rukuni and Anandajayasekeram, 1994, World Bank, 1997). A recent World Bank agricultural sector strategy paper, Rural Development: From Vision to Action, highlights many of these issues along with the objective of targeting a real agricultural growth rate of at least 4 percent annually through improved technology and increased productivity (World Bank, 1997). Within this broad goal, the World Bank identifies more specific objectives aimed at promoting the development impact to: i) increase food production and farm income; ii) make households food, water and energy secure; and iii) restore and maintain the natural resource base. To achieve these objectives the World Bank's agricultural strategy in Africa has identified five key elements largely corresponding to the variables discussed above. These include:

- *Improvement of the policy regime*: creating an enabling policy environment for entrepreneurship and agribusiness development.
- *Technology development and adoption*: promoting technical progress at the firm level and strengthening researcher-extension-farmer linkages.
- *Rural infrastructure*: establishing rural roads, water supply and sanitation, electrification, social services, and improving urban-rural linkages.
- *Empowerment of rural people*: increasing farmer participation in the agricultural and rural development process through political, administrative and fiscal decentralization, participatory project and program development and execution, and by strengthening farmer organizations.
- *Natural resource management*: improving the sustainability of production systems through better management of soils, water, pasture and forests (World Bank, 1997).

This rural economy oriented strategy emphasizes complementarity of decentralized public sector interventions, market-oriented approaches, and popular participation (World Bank, 1997). While many of the growth inducing variables discussed (price and non-price factors) have had an impact on agricultural growth, the mechanisms through which they effect growth sometimes differs. An understanding of these mechanisms will provide a better insight into the growth potential resulting form changes in these variables.

These growth principles have been extensively discussed in the economic growth literature (Barro and Sala-i-Martin, 1995; Durlauf and Quah, 1998). Fafchamps (1998) discusses them in the context of Africa and separates the explanation of growth into a number of functional headings which include elementary, static, dynamic and agglomerate engines (Figure 1.3). This section will briefly summarise these different growth engines in the context of agriculture. A recent text by Eicher and Staatz (1998), includes a detailed discussion on some of these issues.

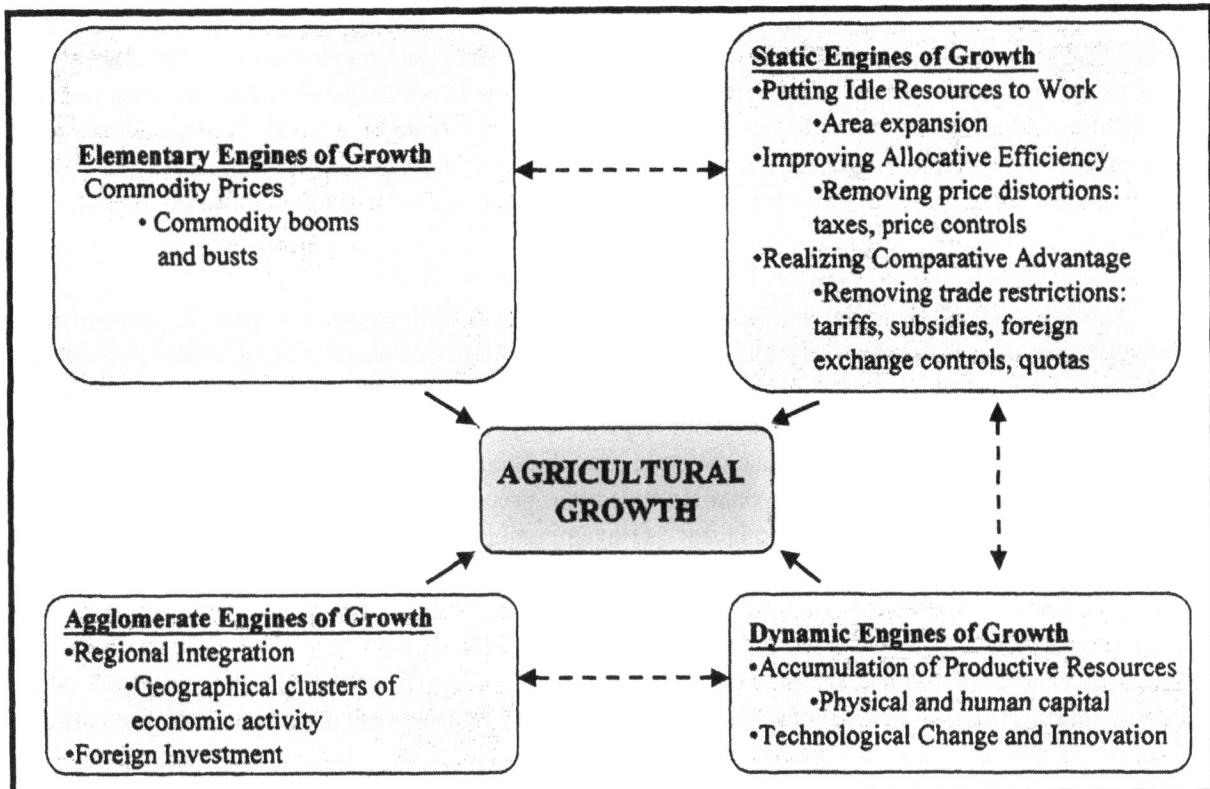

Figure 1.3: Engines of Agricultural Growth.
Source: Adapted from Fafchamps, 1998[10].

Elementary engines of growth: Elementary engines of agricultural growth relate to growth induced by changes in agricultural commodity prices. During commodity price booms many countries in Africa experienced windfall gains. Likewise when commodity prices declined, these countries experienced windfall losses. Gains were particularly evident in the 1970s and the mid-1990s with commodity boom years. Most countries in Africa have an export structure highly dependent on a small number of primary exports. This high level of concentration has resulted in extreme gains or losses. An attempt has been made to reduce these fluctuations and raise world agricultural prices for some commodities using International Commodity Agreements (ICAs) (cocoa, coffee, etc). ICAs, however, have not been able to effectively control production.

Static engines of growth: Static engines of growth relate to growth induced by a change in economic efficiency. These include putting idle resources to work, improving allocative efficiency and realizing comparative advantage. This concept of growth requires a modification of the structure of production to generate growth increases (Fafchamps, 1998). In Sub-Saharan Africa, the first of these, putting idle resources to work has played a significant role in agricultural growth with an expansion in area planted to cereals of 3.4 percent per annum since 1980. This will continue to be an important avenue for growth in Africa, however, this source of growth is finite. Once all resources are fully employed and the production possibilities frontier has been reached, other sources of growth must be found. Improving allocative efficiency offers another source of static growth. It is well known that policies have distorted

---

10 Fafchamps (1998), discusses the applicability of current growth theories to the African context.

agricultural prices (Schiff and Valdes, 1992) and allocative efficiency can only be improved if these distortions are removed. Although it is clear that allocative efficiency can improve social welfare, it is hard to see it as a long term engine of growth. Comparative advantage in production has also not been realised due to distorted policies such as tariffs, subsidies, foreign exchange controls and quotas. These elements of growth have played a large role in the current situation in Africa; indeed significant measures have been taken to remove these inhibiting distortions through structural adjustment programs (World Bank 1994, 1996).

The removal of price distortions has usually been associated with a one time improvement in output through increases in variable factors and has thus been referred to as generating 'static' growth (i.e. after the one off improvement then alternative sources of growth need to be found). This view, however, implicitly assumes that private investment is not responsive to price. Indeed, this is not the case, and there will be some spill-over effects which cause dynamic growth through improved incentives for private sector investment. This investment will be an ongoing economic process stimulating growth. Dynamic growth will also occur through induced innovation effects (Hayami and Ruttan, 1985).

Dynamic engines of growth: Growth induced by an accumulation of productive resources (human, physical and biological) is usually referred to as dynamic growth. As more of these productive resources become available and are used, the production possibilities frontier shifts outwards and output increases. Accumulation theories are implicitly based on the idea that growth is due to more of the same, a process which has limitations for accelerated growth. Further output increases will only be realised with technological change. This was clearly evident in the evolution of mechanised grain harvesting in the United States, evolving from the sickle to the harvester in the mid-1800s. Once the sickle has been adopted (and as a farmer can only use a limited number of sickles), further gains in output (labor productivity) could only be achieved with innovation (development of mechanised harvesting). Similarly, once hybrid seeds have been adopted, further gains will require the development of new varieties[11]. Human capital (education and skills) levels are complementary to capital accumulation as these are required by workers to make effective use of new machinery and new seed varieties (Schultz, 1961).

Agglomerate engines of growth: Agglomeration effects have become increasingly more recognised as a source of additional growth. Regional integration provides a good example of this effect, with several convincing arguments presented in its favour for Africa (Sachs, 1998). The reasoning is that infrastructure is intrinsically a regional affair with market size in individual countries usually being too small to sustain growth and far too small to sustain modern infrastructure by itself. Regional bargaining power is more powerful than that of any single country and a good regional reputation can attract larger amounts of foreign investment. Technology spillovers are also an important element to stimulate regional as well as country specific output and productivity growth.

Agricultural productivity has largely been associated with the dynamic engines of growth, and more especially with the result of technological progress. Improved human capital (health and education) and new technologies (mechanical and biological) are seen to raise land and labour productivity thus stimulating overall agricultural growth. General prescriptions to raising productivity have therefore focused almost exclusively on increasing investment levels in these two areas.

---

11 This is a fairly simplistic argument, as there are many factors involved with the adoption and use of new machines and seed varieties, but it is used simply to illustrate the difference between static and dynamic growth.

EFFICIENCY, TECHNICAL PROGRESS AND PRODUCTIVITY CHANGE IN AFRICA

Recent methodologies allow us to examine this hypothesis by decomposing productivity measures into changes in efficiency and technological progress (Fried *et al*, 1993). Thus productivity growth can be defined as the net change in output due to the change in efficiency and technical change, where the former is understood to be the change in how far an observation is from the frontier of technology and the latter is understood to be shifts in the production frontier (Grosskopf, 1993).

In a world in which inefficiency exists, as is suggested in Africa (and other regions), if this decomposition is ignored, productivity growth no longer, necessarily tells us about technical change. The decomposition is also important from a policy perspective, as a slowdown in productivity growth due to increased inefficiency suggests different policy recommendations to those suggested when the slowdown is due to a lack of technical change. An example is that slow productivity growth due to inefficiency may be due to institutional barriers constraining the diffusion of innovation. In this case, policies to remove these barriers may be more effective in improving productivity than policies directed at innovation.

Several studies have estimated cross-country agricultural efficiency measures (Arnade, 1994; Thirtle *et al*, 1995, Millan and Aldaz, 1998, Fulginiti and Perrin, 1998), which have all been based on data envelopment analysis, employing the Farrell (1957) concept of efficiency (Box A1). Using this same methodology, technical efficiency frontiers were calculated for five regions for 1996/97 and average technical efficiency scores of all countries in each region were calculated (Figure 1.4). The results suggest that Africa is lagging behind other regions in terms of technical efficiency in production. On average, the countries in other regions are closer to their efficiency frontier than countries in Africa, thus reducing this distance could provide a significant future source of agricultural growth.

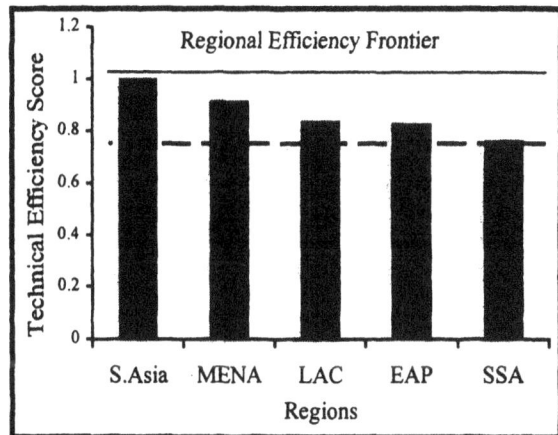

Figure 1.4: Comparison of Technical Efficiency by Region.
Source: Derived using FAO data.

Using a similar methodology, Thirtle *et al* (1995) decompose African agricultural productivity growth into efficiency change and technical progress. They attempt to explain changes in technical progress with agricultural research and development expenditures (R&D), secondary education and machinery (tractors) which all have significant effects. Since R&D is supposed to generate new technology and is complementary to education, these results are not surprising, and tractors do embody new technology. Thirtle *et al* (1995) also find evidence supporting the Boserup/Hayami and Ruttan view of population pressure being causally prior to intensification and land-saving technical change. Changes in efficiency were explained by variables on infrastructure (irrigation and transportation), extension, primary education, the weather and agricultural policies.

Improving efficiency has indeed been an objective of structural adjustment programs. The common philosophy is that more efficient input and output markets, enhanced access to more adequate technology, and more proficient public services together with direct price adjustments improve the output-input price ratio which spurs growth and development (Meerman, 1996). ...'In other words - markets (financial, foreign exchange, factor, and goods), should operate without distortions. Pricing of tradable goods should be at border-parity values and taxes should be neutral. Thus, production is reallocated to higher value output. Hence, the value of output may be expected to expand in the short to medium run because of more efficient allocation of resources, and in the long run because of higher and more efficient investments' (Meerman, 1996, pg. 29). Indeed, this brings us back to the concepts of static and dynamic growth.

GROWTH AND MARKETS

The state of the market has an impact on all the engines of growth highlighted in the previous section (Stiglitz, 1989). Markets can contribute to the development process in two general ways. Firstly, they can provide a way to allocate resources ensuring the highest value production and maximum consumer satisfaction. Secondly, they may stimulate growth by promoting technological innovation and increased supply and demand (dynamic growth) [Scarborough and Kydd, 1992].

Co-ordination and Price Transmission

Following the static and dynamic concepts defined in the previous section, two important ways to achieve higher resource productivity are to increase specialisation in production (an exploitation of absolute and comparative advantage) and technological innovation. The corollary of specialisation is the separation of producers and consumers of particular products into distinct groups. For example, farmers specialising in the production of a certain mix of crops may become dependent on exchange for agricultural inputs, consumer goods, and other agricultural products not produced on their farms possibly including foodstuffs (Timmer, 1997). Specialisation in production therefore requires co-ordination between producers, distributors and consumers or between supply and demand. Markets can provide an important means for this co-ordination.

Growth in the size and spatial spread of the market allows for greater specialisation in production which can lead to increased output from each unit of resource employed. Therefore, a two-way relationship exists between increasing resource productivity and developing adequate systems of exchange. The detailed way in which prices are generated and transmitted through these markets is crucial in determining producer incentives and whether resource allocation leads to maximum production and optimum consumer satisfaction.

Policies, Price Formation and Agricultural Incentives

Prices signals through these markets have often been controlled by governments who defend their interventions as measures required to correct market imperfections/failures. Indeed, another debated topic is the appropriate balance between the state and the market (Williams, 1992).

Several authors highlight the impact of these past government interventions focusing on the extent of agricultural taxation and its implication on agricultural incentives (Schiff and Valdés, 1992,

Herrmann, 1997). The emerging evidence is that past policies have tended to tax agriculture heavily with farmers receiving producer prices significantly lower than the world price equivalent[12]. Indeed, taxation of export commodities by the state, by parastatal bodies, by monopolies or via corruption or holdups such as police barricades has been a common form of extraction from agriculture (Binswanger and Deininger, 1997). These means of extraction undermined the incentives to produce and invest, and tend to have a much higher deadweight loss than the more modern forms of taxation. Nevertheless, the taxation would not have been as crippling to agricultural growth if some of the surplus was invested into public services and infrastructure (Newbery, 1992). Indeed East Asia provides some 20th century examples of high extraction from rural areas combined with significant public investment and growth (Karshens, 1998). The increase in the flow of public funds to agriculture significantly enhances agricultural income. Government spending on agricultural research and education raises the productivity of factors (dynamic engines). Investments in transportation reduces internal marketing costs, increases the net income accruing to agriculture, and opens up new areas for production.

These high levels of taxation were quantified by Schiff and Valdés (1992) using data from eighteen developing countries. They show that Sub-Saharan African countries (Côte d'Ivoire, Ghana and Zambia) imposed the most severe taxation (both explicit and implicit) on agriculture, ranging from 46-59 percent. The direct tax on agriculture in these African countries was similar to the level of the indirect tax, implying that agricultural pricing policies taxed agriculture about as much as the implicit tax resulting from industrial protection and macroeconomic policies. This differs markedly from their findings in the other developing countries where the implicit tax was nearly three times that of the direct tax. Herrmann (1997) conducted a similar study which focuses on individual crops (coffee, wheat and rice). He also finds significant policy biases against agriculture, these being more excessive for the export crops (coffee) than the food crops (rice and wheat). The favorable agricultural policies for food crops were often shown to be offset by distorted macroeconomic policies with a resulting decline in the real producer price. Both of these studies use pre-1985 data, which limits their use in identifying the current distortions facing farmers in Africa.

Progress in reducing these high level of taxation was made by the end of the 1980s. Structural adjustment programs with macroeconomic, trade and sector reforms, had been widely adopted. The intended impact on the agricultural sector was to increase the payoffs to small-holder farmers, primarily from an expansion in the production of export crops, thereby accelerating rural growth and poverty reduction. Price incentives improved in many countries as shown in the 1994 World Bank study - *Adjustment in Africa.* Of the twenty seven countries analyzed, ten experienced an increase in the real producer price of export crops, while seventeen experienced a decline. The explanation provided for the large declines was the fall in world prices, together with the countrys' inability to reduce both explicit and implicit taxation simultaneously. Consequently, the benefits from the reduction in one were normally eroded by the losses from the increase in the other.

---

[12] Much of the focus on price relationships in the literature has been on the trends in the ratio of primary prices to manufacturing prices. The argument is that, in the long run, this ratio moves downwards (Prebisch, 1950 and Singer, 1950). This, together with the perception that agriculture was impervious to price incentives and that there were significant benefits from reducing the reliance on the importation of manufactured goods led to a belief that industrialization was the key to rapid growth. Following these perceptions, the conventional wisdom was to tax the primary sectors, such as agriculture, to provide resources to build a modern industrial sector.

A more recent study (World Bank, 1996) which examined the change in real producer prices over the decade 1985 to 1995 for several major export crops (coffee, cotton, cocoa, cashew) in five countries (Benin, Burkina Faso, Côte d'Ivoire, Mali, Mozambique and Uganda) suggests a similar result. Despite macroeconomic and agricultural reforms, several countries experienced stagnant real producer prices. The study revealed a lack of pass-through of international prices to domestic prices suggesting that there are some remaining impediments preventing farmers from enjoying the full benefits of these recent policy changes. These impediments can be explained by both price and non-price factors and the linkages between the two.

COMPLEMENTARITIES BETWEEN PRICE AND NON-PRICE FACTORS

Indeed, there are large complementarities between price and non-price factors. This was clearly illustrated empirically by Schiff and Montenegro (1997), using data from eighteen developing countries. As mentioned in the previous section, these non-price factors include investments in new technologies, rural infrastructure and human capital. The complementarities suggest that the removal of price distortions through macroeconomic and sector reforms will only have limited impact on farmer response if market infrastructure, institutions and support services are undeveloped (Binswanger, 1989, Tshibaka, 1997). In these cases transaction costs will be high resulting in high input costs and low output prices, thus some of these non-price factors implicitly effect prices. Conversely, the removal of structural and institutional constraints alone will only have a limited impact on the farmer's response if price distortions continue to be significant.

This study will highlight some of the complementarities, in particular between prices and rural infrastructure with a brief discussion on the institutional framework in African countries. Some aspects of technology will be discussed but they will not be the primary focus of this particular study. Indeed the challenge of improving agricultural incentives is highlighted by the complex nexus among the structures of the broader agricultural economy (marketing systems, transportation costs, infrastructure), the macroeconomic environment (trade, exchange rates, institutional framework) and the political arena in which agriculture operates. This study will take stock of current policies as well as their recent evolution to update knowledge and to generate discussion to help develop a stronger consensus on an appropriate policy stance that will improve agricultural incentives and raise average agricultural growth rates and farm incomes.

OUTLINE OF THIS WORK

The next section examines the external market environment focusing on the recent world price trends and volatility in commodity markets. The external terms of trade for African countries are calculated. Access of African countries to OECD markets is also discussed. The remainder of the study examines the internal market environment. Chapter three assesses the current macroeconomic policy stance of African countries and discusses the evolution of fiscal, monetary and exchange rate policies since 1990. The linkages between these macroeconomic policies and agricultural prices and income are highlighted through country case studies.

Chapter four focuses on export crop market reforms and price trends. The current state of the market and the recent evolution of marketing policies are reviewed. Exchange rate pass-through to producer prices are examined together with complementarities between prices and rural infrastructure.

Food crop prices and markets are examined in chapter five. Maize producer and consumer price trends are reviewed. Private trader emergence into liberalized food markets is examined, identifying some of the barriers to entry.

The study only examines one strategic agricultural input in chapter five, namely fertilizer which is the largest purchased input in Africa agriculture. The chapter examines the fertilizer market with a focus on fertilizer price trends, exchange rate pass-through, transport infrastructure, private sector participation, trade restrictions and fertilizer aid.

Chapter seven draws up several policy challenges facing Africa, elaborating on a strategy to remove market and price barriers to improve agriculture incentives. Country profiles of price policies for sixteen Sub-Saharan African countries are presented in chapter eight.

PART II

THE EXTERNAL MARKET ENVIRONMENT

2. WORLD PRICES AND MARKET ACCESS

The real world price of the major agricultural exports from Africa has declined over the past several decades. These declining trends have had a significant impact on African economies as agriculture is the dominant sector in most of these countries, accounting for 40 percent of total exports from the continent. Within agriculture, nine crops account for about 70 percent of total exports and this share has remained relatively constant since the 1970s.

Table 2.1: Sub-Saharan Africa's Commodity Export Earnings as a Percentage of Total Agricultural Exports.

Crop	1970-79	1980-89	1990-97	Growth Rate (1970-97)
Bananas	0.7	0.5	1.3	2.4
Cocoa	20.6	22.1	18.5	-0.2
Coffee	24.7	25.9	14.7	-2.1
Cotton	9.2	8.6	11.9	0.9
Groundnuts	2.4	0.6	0.3	-8.5
Rubber	1.7	2.1	2.4	1.7
Sugar	5.6	6.8	8.5	0.7
Tea	2.5	3.6	4.8	3.3
Tobacco	3.1	4.8	8.8	4.7
% of Total Agric. Exports	70.4	74.9	71.2	-

Source: FAO Trade statistics. These calculation exclude South Africa. Including South Africa reduces the last row percentages to 62, 66 and 60 for the consecutive periods.

Although the aggregate share has been fairly stagnant, the shares of several individual crops have changed significantly. Cocoa and coffee are the dominant crops and accounted for over 40 percent of the continents agricultural exports between 1970 and 1990. Since 1970, these two crops have experienced a declining share with an annual decline of -0.2 and -2.1 percent for cocoa and coffee respectively. Cotton has grown in importance, its export share increased from 8.6 percent between 1980-89 to 11.9 percent between 1990-97. Conversely, the export share of groundnuts declined from 2.4 percent in 1970-79 to 0.6 percent in 1980-89. This decline was largely caused by aflotoxins, which reduced external demand for African groundnuts on international markets. The continual commodity dependence observed in Table 2.1 makes many African economies susceptible to serve terms of trade fluctuation from commodity price shocks. The impact of these shocks can be widespread, affecting exchange rates, the cost of debt, government revenue, and private producers, processors, traders and consumers.

Table 2.2: Sub-Saharan Africa's Share of World Trade.

Country	Levels		Averages			Annual Growth Rates
	1970	1997	1970-79	1980-89	1990-97	1970-1997 (%)
Bananas	7	4	6	3	4	-3.3
Cocoa	68	41	59	45	40	-2.0
Coffee	26	12	27	22	14	-3.1
Cotton	17	14	13	11	12	-0.2
Groundnuts	68	7	44	11	5	-10.2
Rubber	8	7	6	6	5	-0.5
Sugar	9	7	8	7	8	-0.2
Tea	13	21	15	15	19	1.3
Tobacco	8	12	8	9	12	1.7

Source: FAO data.

Although SSA has experienced a loss in its international market share of agricultural exports, with a decline of 27, 14 and 61 percent for cocoa, coffee and groundnuts respectively since 1970, the most recent trends have been more favorable for many crops (Table 2.2). The export share for five of the nine crops (bananas, cotton, sugar, tea and tobacco) has increased since the 1990s.

WORLD COMMODITY MARKETS AND PRICE TRENDS

Recent studies (Cuddington *et al*, 1993; Bleaney and Greenaway, 1993; Reinhart and Wickham, 1994) have highlighted the long-term downward trends in the international terms of trade for primary commodity exports relative to manufactured goods[13]. Reinhart and Wickham (1994) and Sapsford and Singer (1998) suggest that prices are not only declining but their variance (volatility) is increasing. They find that the major part of the recent weakness of commodity prices is associated with secular (permanent) as opposed to the cyclical (temporary) trends.

Secular Trends

Secular trends are permanent (irreversible) trends in nature whose importance varies considerably across commodity groupings. Reinhart and Wickham (1994) suggest that about 85 percent of the variance of world beverage prices is due to permanent shocks, while for metals, these permanent shocks only account for 30 percent of the price variance. They go on to highlight the factors affecting these secular (permanent) price declines which include agricultural policies in industrial countries, technological innovation and to some extent, the breakdown of international commodity agreements.

Subsidies afforded to agriculture by industrial countries have induced a large expansion in supply, particularly evident in the 1980s, thus contributing to world price declines. The development and adoption of new technology has significantly increased average yields, raising world output and reducing world prices. The breakdown of the ICAs for sugar (1984), cocoa (1988), coffee (1989) caused a surge in supply and a large 'temporary' downward price spike. Reinhart and Wickham (1994) suggest that the ICA breakdowns are also likely to have had 'permanent effects', as exporters' behavior adjusts to the more competitive regimes.

Cyclical Trends

Cyclical price trends are temporary in nature. Typical causes of these price fluctuations include the effect of weather (drought, frosts and floods), as well as recessions in industrial countries. For example, slower demand in recent years from Russia and the effects of the Asian crisis have all caused a temporary decline in the world price of a number of agricultural commodities. The nature of the cyclical trends determines the differing degrees of trend persistence.

World Price Volatility

In more recent years, the volatility of world prices has also received increasing attention. Reinhart and Wickham (1994) show that between 1957 and 1993, commodity price volatility increased around the declining price trends. Using a similar methodology, with a longer time period (1960 to

---

[13] The Prebisch-Singer hypothesis.

1998), with more focus on crops produced by Sub-Saharan Africa, Table 2.3 compares the means and variance of commodity prices over time. There are several notable features. First, the average prices in the latter periods are lower, consistent with the negative trend. Second, although there is a sustained and sharp increase in the variance of several world commodity prices between the 1960s and 1980s, this variance declines in the 1990s for all commodities except bananas and tobacco. (see Figure A1 in the appendix).

Table 2.3: Mean and Variance of World Commodity Prices.

Commodity	Statistic	1960-70	1970-80	1980-90	1990-98
Bananas	Mean	6.547	6.275	6.273	6.106
	Variance	0.006	0.009	0.008	0.017
	CV	1.191	1.546	1.466	2.157
Cocoa	Mean	5.541	5.902	5.474	4.829
	Variance	0.046	0.165	0.142	0.021
	CV	3.877	6.890	6.886	3.012
Coffee	Mean	5.713	6.087	5.687	4.980
	Variance	0.037	0.134	0.185	0.125
	CV	3.358	6.019	7.557	7.086
Cotton	Mean	5.670	5.694	5.317	5.023
	Variance	0.006	0.026	0.066	0.024
	CV	1.375	2.823	4.824	3.066
Groundnuts	Mean	6.030	5.994	5.505	5.056
	Variance	0.003	0.066	0.057	0.033
	CV	0.974	4.271	4.328	3.592
Rubber	Mean	5.534	5.226	4.962	4.694
	Variance	0.054	0.032	0.052	0.037
	CV	4.202	3.428		4.116
Sugar	Mean	3.437	3.995		3.095
	Variance	0.239	0.302	0.304	0.019
	CV	14.222	13.750	16.914	4.413
Tea	Mean	6.095	5.687	5.463	5.122
	Variance	0.025	0.028	0.067	0.034
	CV	2.578	2.940	4.733	3.616
Tobacco	Mean	8.722	8.263	8.125	7.970
	Variance	0.040	0.015	0.016	0.028
	CV	2.282	1.480	1.541	2.089

Source: World Bank data, Price were in logarithms. CV is the coefficient of variation.

The sharpest rise in volatility appears to have taken place during the early 1970s following the breakdown of the Bretton Woods exchange system after the fist oil shock. Volatility peaked during the 1980s and has since declined in seven of the nine commodities presented in Table 2.3. The decline in volatility is apparent when international commodity agreements have largely collapsed and where the trade regimes across countries have become less distorted with the implementation of the Uruguay commitments.

Over the last several decades, African countries have adopted a number of mechanisms to deal with price risk. However, the design and feasibility of these mechanisms (such stabilization and hedging strategies) depend very much on the nature of the shock. The response to cyclical volatility in commodity prices, raises issues different from those emanating from secular changes. Income stabilization policies and hedging are useful in dealing only with temporary and, preferably, short-lived shocks. The ability to stabilize prices depends on the persistence of the shock and the duration of the cycles. Permanent shocks may require adjustment (export diversification, [Sapsford and Singer, 1998]).

In Africa, although many marketing boards and stabilization schemes were set up, their record of effectively stabilizing prices and incomes has been mixed (this will be discussed in more details in chapter 5). A number of commodity exchanges have developed in Africa (Zimbabwe and South Africa) to manage price risk in the more open markets. What has become apparent is that developing countries, in general, have faced certain problems in accessing market-based risk management instruments. Larson *et at* (1998) highlight some of these problems and offer several solutions. They suggest that the problems associated with government entities center on the capacity and incentives to develop and execute risk management strategies while for private entities, lack of knowledge and understanding, inadequate legal, regulatory and institutional environments limit access to current markets.

Diversification strategies to cope with permanent price shocks have included producing more non-traditional export crops such as citrus fruits, cut flowers, spices and animal products (e.g. Kenya and Zimbabwe). A comparison of the price trends of these commodities with the traditional agricultural exports (Table 2.4) shows that between 1961 and 1997 the world price of traditional exports declined at a rate over three times faster than that of the non-traditional exports. However, from 1990-1997, with the boom in traditional export crops, particularly in coffee, the trends between these two groups have been reversed. The large decline in the world vanilla price in the 1990s can largely be attributed to the introduction of vanilla substitutes on world markets.

Table 2.4: World Price Trends of Traditional and Non-Traditional Exports.

Commodity	Growth Rate 1961-97	Growth Rate 1990-97	Commodity	Growth Rate 1961-97	Growth Rate 1990-97
Non-Traditional Exports			*Traditional Exports*		
Citrus Fruit	0.8	3.2	Bananas	-1.4	-3.6
Pineapples	-0.3	3.2	Cocoa	-2.2	2.7
Pulses	-2.3	-2.2	Coffee	-1.8	11.2
Live Animals	-1.9	-2.4	Cotton	-2.4	0.5
Beef and Veal	-0.6	-3.3	Groundnuts	-3.3	0.8
Hides and Skins	-1.6	-0.5	Rubber	-2.7	5.0
Cashew Nuts	0.5	-1.9	Sugar	-2.0	0.8
Vanilla	-0.3	-11.2	Tobacco	-2.5	-3.2
			Tea	-3.3	-2.0
Mean	-0.7		*Mean*	-2.4	

Source: World Bank, FAO.
Note: To derive a world price for the non-traditional exports, the value of world exports was divided by the quantity of world exports. Data were derived from the FAO trade statistics.

Price Trends During the 1990s

The recent commodity boom, which reached its peak in 1995-96, has provided temporary relief to the long-term downward movement in commodity prices. Although temporary (cyclical), this boom provided welcomed benefits to farmers (Table 2.5 and Figure 2.1 to 2.4). Between 1990-91 and 1996-

97, the price of export crops increased at a greater rate than that of both the price of grains and the price of fertilizer. However, this positive trend is unlikely to continue as shown by the forecasts in Table 2.5.

Table 2.5: World Commodity Price Trends.

Percentage Change in the Real World Price (Deflated by the MUV)											
1990-91 to 1996-97						**1996-97 to 2010 (Forecast)**					
SSA Export Crops		**Food Crops**		**Fertilizer**		**SSA Export Crops**		**Food Crops**		**Fertilizer**	
Bananas	-19	Wheat	+18	Phosphate	-12	Bananas	-17	Wheat	-26	Phosphate	-16
Cocoa	+14	Rice	+4	Urea	-3	Cocoa	+8	Rice	-12	Urea	-17
Coffee	+43	Maize	+19	TSP	+20	Coffee	-18	Maize	-25	TSP	-32
Cotton	-8			DAP	+9	Cotton	-15			DAP	-26
Groundnuts	+18					Groundnuts	-13				
Rubber	+30					Rubber	-17				
Sugar	-1					Sugar	-3				
Tea	-8					Tea	-18				
Tobacco	-13					Tobacco	-21				
SSA Export Crops	**+20**	**Grains**	**+14**	**Fertilizer**	**+8**	**SSA Export Crops**	**-7**	**Grains**	**-20**	**Fertilizer**	**-27**

Source: World Bank.

African Export Crops

Several export crops, such as coffee, rubber, cocoa and groundnuts experienced large price increases between 1990 and 1995-96. The low coffee prices in the early 1990s were caused by the weak demand from traditional markets (Soviet Union). The large world stocks created by high prices in the mid-1980s led to the collapse of the world coffee organization's buffer stock and price support scheme. Between 1990 and 1994 coffee prices increased by over 150 percent. The 20 percent aggregate increase in the real price of SSA export crops is unlikely to continue as the effects of the Asian crisis are felt on the world markets.

Grain Crops

A key indicator of the world grain prices is the stock levels in the five largest grain-exporting countries (USA, EU, Canada, Australia and Argentina). Low world grain stocks coupled with the poor US harvest resulted in large real world price increases in 1995-96. The subsequent decline in 1997 was the result of the large supply response to the higher 1995/96 prices and weak world demand.

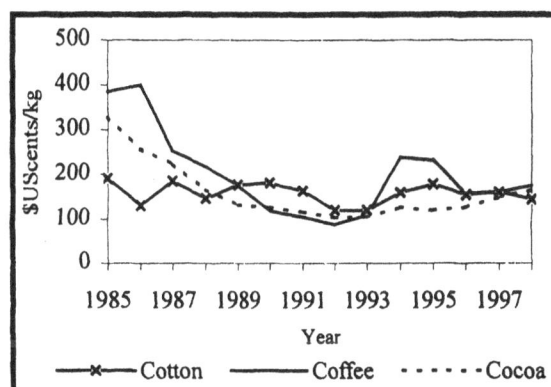

Figure 2.1: Trends in the World Price of Cotton, Coffee and Cocoa, 1990 Constant Prices, US cents/kg. *Source:* World Bank data.

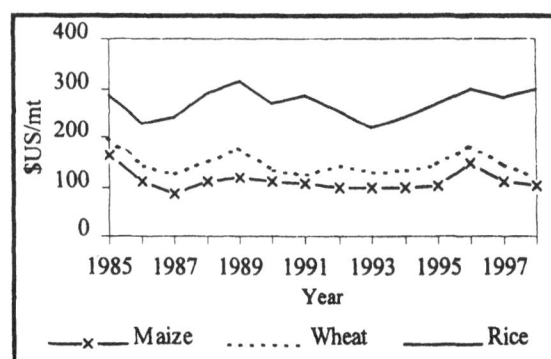

Figure 2.2: Trends in the World Price of Maize, Wheat and Rice, 1990 Constant Prices, $US/mt. *Source:* World Bank data.

Fertilizers

The low fertilizer prices in the early 1990s were due to a glut in the fertilizer market which caused several fertilizer plants to close. The subsequent price increases are mostly due to the tight supply that fertilizer manufacturers have been enjoying throughout the past several year, as well as the sharp increase in fertilizer consumption in China, Vietnam and Pakistan (Gerner *et al*, 1996). Fertilizer prices are forecast to decline by more than the declines predicted in export crops and grain crops (Table 2.5).

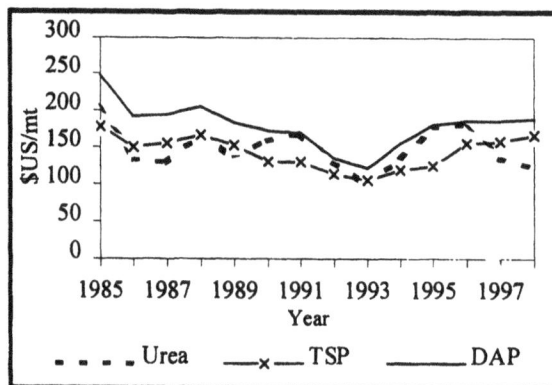

Figure 2.3: Trends in the World Price of Urea, DAP and TSP, 1990 Constant Prices, $US/mt.
Source: World Bank data.

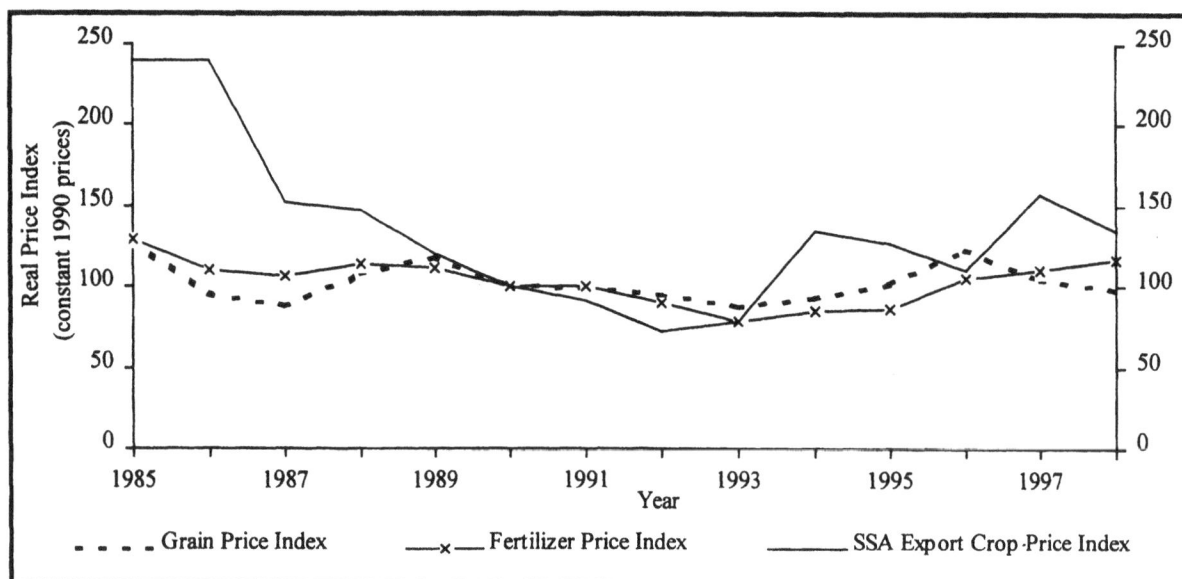

Figure 2.4: World Commodity Price Trends, 1985-1997.
Source: World Bank.

EXTERNAL TERMS OF TRADE

The favorable world price trends in recent years have improved the external terms of trade for agriculture (Table 2.6). External barter terms of trade for agriculture[14] have improved at a greater rate than the external barter terms of trade for the economy[15] of many African countries. The improved external terms of trade for agriculture suggest that the purchasing power from exporting agricultural commodities is increasing relative to food imports. Similar trends are derived if the manufacturing unit value (MUV) is used as the denominator.

---

[14] Measured as the ratio of the price of agricultural exports to the price of food imports (Table A2).
[15] Measured as the ratio of the unit value of total exports to the unit value of total imports for the economy

In most African countries examined, the external net income terms of trade have increased at a greater rate than that of the barter terms of trade (see Appendix for calculation, Table A2). In three of the 16 countries analyzed (Zimbabwe, Mozambique, and Mali) the external barter terms of trade have deteriorated while the external net income terms of trade have improved because of the larger percentage increase in the volume of agricultural commodities exported. The reverse is true in other cases (Madagascar) where the increase in the external barter terms of trade has been eroded by the decline in volumes exported. In some cases, the terms of trade improvement have been phenomenal (Uganda for instance had a growth rate of 8 percent and 22 recent per annum in its barter and net income term of trade respectively).

The last column in Table 2.6, showing the change (growth rate) of the crop export price to the world fertilizer price, indicates large contrasts across countries for the 1990s. The growth rate in ten countries (Burkina Faso, Cameroon, Côte d'Ivoire, Nigeria, South Africa, Tanzania, Madagascar, Ethiopia, Kenya and Uganda) is positive, with high growth rates in Uganda, while it is negative in six countries (Mali, Niger, Ghana, Malawi, Mozambique and Zimbabwe), with large negative values for Niger, Malawi and Zimbabwe. This indicates that the contrasts in prices result from differing compositions of export commodities across countries with some countries experiencing a greater decline in world prices of their agricultural commodity exports. The sign on this ratio is the same as that in the external barter terms of trade for agriculture, and in some cases, the variations are larger.

Table 2.6: External Terms of Trade in Sub-Saharan Africa, 1989/90-1996.

Country	Barter Terms of Trade (Agriculture) Growth Rate p.a.	Net Income Terms of Trade (Agriculture) Growth Rate p.a.	Barter Terms of Trade (Economy) Growth Rate p.a.	Agricultural Export Price/World Fertilizer Price Growth Rate p.a.
SSA	**1.63**	**4.21**	**-0.87**	-
CFA countries				
Burkina Faso	4.2	0.9	1.4	4.1
Cameroon	11.9	12.9	-4.4	3.4
Côte d'Ivoire	3.9	6.8	3.2	3.9
Mali	-1.4	1.4	-1.8	-3.1
Niger	-5.1	-10.2	-4.1	-8.2
Non-CFA countries				
Ghana	-5.9	-2.7	0.13	-1.3
Nigeria	7.0	10.5	-4.5	9.0
Malawi	-2.3	-0.8	-3.1	-5.6
Mozambique	-0.8	6.6	-	-1.2
South Africa	4.6	0.7	0.6	4.5
Tanzania	1.9	8.7	-0.8	3.1
Zimbabwe	-1.9	9.6	-1.5	-3.8
Madagascar	7.4	3.4	1.6	5.3
Ethiopia	0.7	10.9	-1.7	5.1
Kenya	1.4	3.4	7.2	1.7
Uganda	8.3	22.2	5.3	10.9

Source: FAO, World Bank and IMF data. The exports and imports used to calculate the terms of trade are shown in the appendix. The largest countries in the region based on agricultural value added (with data available) were used for this analysis.

The improved external terms of trade for agriculture in the 1990s has been facilitated by real effective exchange rate depreciations. CFA countries' terms of trade improved more than those of the non-CFA countries due to the more favorable world prices for crops such as coffee and cocoa, which are largely produced by the CFA countries. The trend in net income terms of trade shows a much larger improvement for the non-CFA countries, a phenomenon which will be analyzed later.

ACCESS TO OECD MARKETS

Another component of the external market environment is Africa's access to OECD markets. Several external factors have had an influence, namely high external transportation costs, import barriers and phytosanitary requirements.

External transportation costs: Several African countries receive preferential trade agreements from the OECD (ie: sugar in Mauritius, beef in Botswana) which were received under the OECD's Generalized System of Preferences (GSP) schemes, and through the European Union Lomé Convention. Amjadi and Yeats (1995), however, show that these preferential tariffs were often offset by the higher than average nominal freight costs on Africa's exports.

Import barriers: In recent years, import tariffs do not appear to have been especially limiting to African export. Ng and Yeats (1996), in fact, show that African countries faced average tariffs below those paid by other exporters. Their analysis concludes that there is no evidence "that OECD tariffs caused the general loss of competitive position reflected in Africa's declining market shares" (pg. 20). Their analysis focused on all traded goods and not just agriculture.

They argue that non-tariff barriers have not been significantly biased against African countries when all trade items are considered. The coverage ratio for SSA countries was lower than that of other developing countries but higher than for intra-OECD trade. However, Sub-Saharan Africa faced a higher incidence of NTBs (Non-tariff barriers) on 'all food items' than other developing countries (Figure 2.5). Coffee exports are subject to quantitative controls imposed under the International Coffee Agreement and special taxes are also applied to coffee imports in several European markets (Ng and Yeats, 1996).

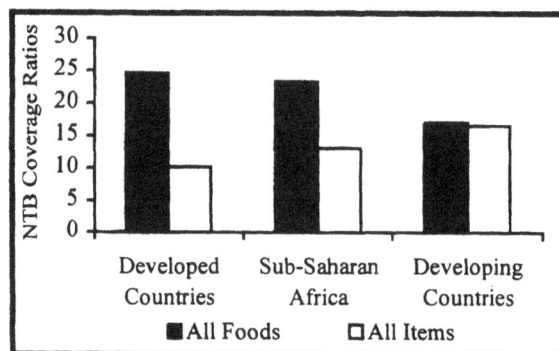

Figure 2.5: Non-Tariff Measure Coverage Ratio for OECD Imports from Developed, Developing and Sub-Saharan African Countries.
Source: Ng and Yeats, 1996.

Under the Uruguay Round Agreements, defined NTBs are being converted into tariff equivalents. These NTBs include quantitative import restrictions, variable import levies, minimum import prices, discretionary import licensing and non-tariff measures maintained through state trading enterprises (Ingco and Townsend, 1998). A special treatment clause (food safety and environmental protection) was introduced to allow certain countries to postpone application of tariffication to sensitive commodities (such as staple products from Africa). Under certain conditions this clause allows the maintenance of import restrictions up to end of the implementation period (Tshibaka *et al*, 1998). There were some concerns that the Uruguay Round may result in market losses for Africa due to the erosion in the value of preferences in its export markets as overall cuts in tariffs reduce the value of the preferences (Davenport *et al*, 1994). Currently, further studies are being conducted to assess constraints to OECD market access (Ingco *et al*, 1999).

Sanitary and phytosanitary requirements: Sanitary and phytosanitary standards remain a barrier to expand trade into foreign markets, but they are conditions that African countries must satisfy. As trade participation grows with reduced trade barriers and development, a country's ability to meet and apply sanitary and phytosanitary standards will become more important for markets access. Applying such standards means building effective system to control or eradicate plant and animal diseases and to ensure the safety of exports. However, the abuse of these conditions by OECD countries as market barriers needs to be closely monitored by the international community and the WTO.

Several domestic factors also limit market access. These include low comparative advantage and efficiency in farm production and high transportation and transaction costs. Some of these issues will be examined in the following chapters.

SUMMARY AND CONCLUSIONS

The long-term (secular) world price trends towards lower real commodity prices looks firmly set in place. These trends suggest that Africa must produce agricultural commodities at lower cost (adopt new technologies) or else its position in world markets will continue to erode. The short-term (cyclical) price variations remain high, although some commodities show a decline since 1990 (sugar, cocoa, cotton and rubber). In order to cope with these price variations, African countries need to improve their access to market based risk management instruments.

From 1990 through 1997, real world commodity prices increased, improving the external market environment for African agriculture.

- Between 1990 and 1997, the real world price of SSA's agricultural exports increased by 20 percent, grain crop prices increased by 14 percent and fertilizer prices increased by 8 percent.
- These trends resulted in a 1.6 percent annual improvement (growth) in the external barter terms of trade and a 4.2 percent annual improvement (growth) in the external net income terms of trade.

The recovery of real world commodity prices for traditional African exports was temporary with a significant price downturn in 1998/99. The long term downward trend in world prices for the non-traditional exports such as cut flowers, citrus fruit, deciduous fruit, live animals and animal products appears to be less severe than the trends in the price of traditional export crops.

- Recent events in the world economy, particularly the economic slow-down of Russia and Asia, are likely to have a considerable temporary negative effect on agricultural commodity prices.
- African countries access to OECD markets varies significantly across crops. High external transportation costs and sanitary and phytosanitary standards remain barriers to expand trade into foreign markets. Unit transport costs could be reduced with larger product shipments. This could be achieved if African countries shared cargo space on ships. This closer collaboration needs to be encouraged through greater regional integration. In order to meet the sanitary and phytosanitary

requirements, African countries will need to build an effective system to control or eradicate plant and animal diseases and to ensure the safety of exports.

PART III

THE INTERNAL MARKET ENVIRONMENT

3. THE MACROECONOMIC POLICY ENVIRONMENT

Since Schuh's (1974) seminal work on macroeconomic policy effects on agriculture, the importance that the macroeconomy can play in generating (depressing) agricultural growth is increasingly recognized. A considerable literature exists and shows that macroeconomic variables do have an important impact on agricultural incentives. Recent work by Schiff and Valdés (1998) provides an excellent summary of these effects. The focus of many African studies on the linkages of the macroeconomy to agriculture has been on the impact of exchange rates (Tshibaka, 1993; Jaeger and Humphreys, 1988, Elbadawi, 1992, Oyejide, 1996, Cleaver, 1985), interest rates and the money supply (Hassan, 1996, Dercon, 1993). In some countries, macroeconomic policy has been shown to have a larger effect on agriculture than sectoral policy (Schiff and Valdés, 1992).

The policies pursued by many African countries in the 1970s and early 1980s were consistent with their inward-looking industrialization policy, with the state having a prominent role in regulating economic activity. Public enterprises were created, regulations enacted to control prices, restrict trade and allocate foreign exchange in pursuit of social goals. Overvalued exchange rates and large and prolonged budget deficits undermined the macroeconomic stability needed for long-term growth(World Bank, 1994)[16]. In Africa, this macroeconomic instability was more influenced by domestic than external shocks (Hoffmaister *et al*, 1997).

Faced with severe macroeconomic problems, such as falling export earnings, worsening balance of payments, mounting debts, and declining economic growth, many countries in Sub-Saharan Africa adopted structural adjustment programs. The programs encompassed a wide range of policy reforms starting with macroeconomic stabilization to correct macroeconomic imbalances. During the 1980s many SSA countries managed to improve their macroeconomic policies and increase their international competitiveness. In a recent study of twenty-six African countries, fifteen made advances in macroeconomic policy during the 1980s, while eleven suffered a deterioration (World Bank 1994). Even though these policies had significantly improved between 1980 and 1990-91, fiscal balances were still fragile, inflation was above international levels, and the parallel market premium for foreign exchange had not been eliminated.

This section will describe the macro-economic environment in Africa and establish the current macro-economic policy stance for a sample of African countries. A set of African countries similar to those used in the *Adjustment in Africa* study will be examined and will provide a useful comparative analysis. Trade policy will not be included in the explicit assessment of overall macroeconomic policy but will be included in the assessment of export crop, food crop and fertilizer policy in chapters four, five and six respectively. While the measures of the macroeconomic policy stance used in the *Adjustment in Africa* study (World Bank, 1994) are open to debate, they provide a useful approximation. The same measures for the fiscal, monetary and exchange rate policy stance will be used in this study (Table A3).

---

[16] There is extensive evidence that improved macroeconomic and sectoral policies are associated with higher growth in Africa (Easterly, 1995, World Bank, 1994, Sachs and Warner, 1998, Jaeger, 1992, Tshibaka, 1993, Valdés, 1992, Binswanger, 1992 and Cleaver, 1985 – this is by no means an exhaustive list).

FISCAL POLICY

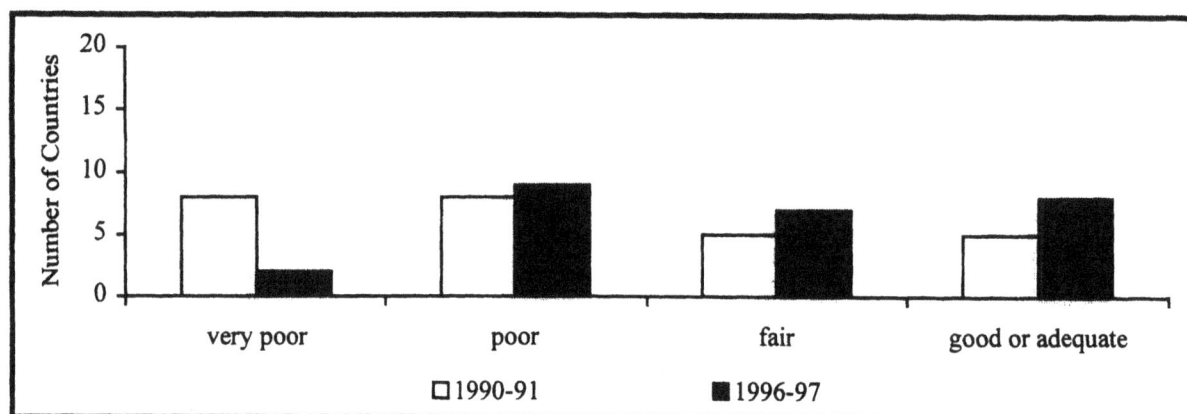

Figure 3.1: Comparison of the Fiscal Policy Stance in Sub-Saharan Africa between 1990-91 and 1996-97.
Source: Table 3.1.

Table 3.1: Fiscal Policy Stance.

Policy Stance*	1990-91	1996-97
Good or Adequate	The Gambia Ghana Mauritania Senegal Tanzania	Benin Cameroon Congo, Rep Gabon Mauritania Niger Nigeria Senegal
Fair	Burundi Burkina Faso Gabon Malawi Togo	Burkina Faso Central African Republic Côte d'Ivoire Kenya Mali Tanzania Uganda
Poor	Benin Central African Republic Kenya Madagascar Mali Nigeria Rwanda Uganda	Burundi Madagascar Malawi Mozambique Rwanda Sierra Leone Togo Zambia
Very poor	Cameroon Congo, Rep Côte d'Ivoire Mozambique Niger Sierra Leone Zambia Zimbabwe	Ghana The Gambia Zimbabwe

* See Appendix for definitions
Source: Adjustment in Africa and Table A3.

Fiscal policies have improved significantly since 1990-91. In 1996-97, eight countries had a 'good or adequate' policy stance (a budget deficit, including grants, of less than 1.5 percent of GDP); this represented an improvement compared with five countries in 1990-91 (Table 3.1, Figure 3.1). The number of countries with 'very poor' policies was also reduced from eight in 1990-91 to three (Ghana, The Gambia and Zimbabwe) in 1996-97 (a 'very poor' fiscal policy means a deficit greater than 7 percent of GDP). While a low fiscal deficit is not a sufficient condition of good policy, it is certainly a necessary component, and high budget deficits, if sustained, are likely to be inflationary in the longer term (World Bank, 1994). Although fiscal policies have improved, a high level of instability remains. For example, Ghana declined from a 'good or adequate' fiscal policy in 1990-91 to a very poor policy in 1996-97. A regional comparison suggests that the CFA countries appear to have improved their fiscal policy stance more than the non-CFA countries since 1990-91. A contributing factor to this may be the 50 percent devaluation of the CFA franc in 1994 that provided higher domestic export revenues. Although there has been a general improvement in the fiscal policy stance since 1990, eleven countries still had a 'poor' or 'very poor' stance in 1996-97.

MONETARY POLICY

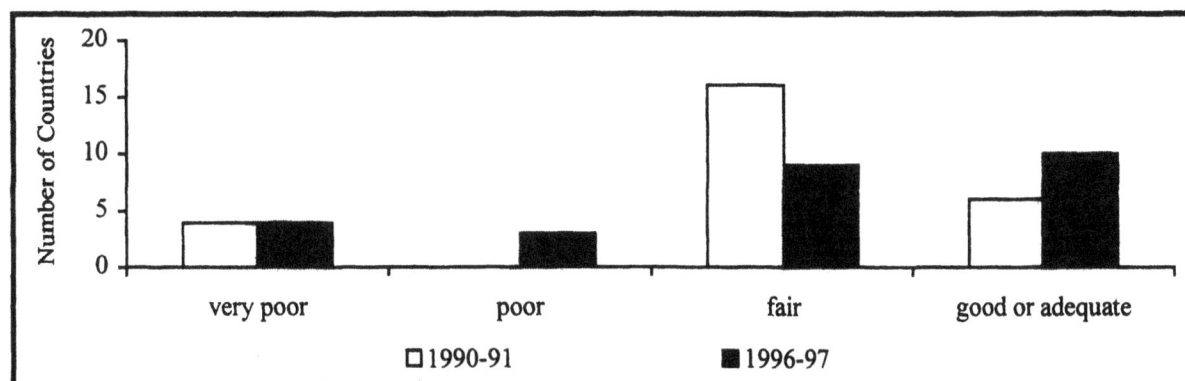

Figure 3.2: Comparison of the Monetary Policy Stance in Sub-Saharan Africa between 1990-91 and 1996-97.
Source: Table 3.2.

Table 3.2: Monetary Policy Stance.

Policy Stance*	1990-91	1996-97
Good or Adequate	Burkina Faso Burundi Central African Republic Côte d'Ivoire Gabon Mali	Benin Cameroon Central African Republic Côte d'Ivoire Gabon Mali Mauritania Niger Senegal Uganda
Fair	Benin Cameroon Congo, Rep The Gambia Ghana Kenya Madagascar Malawi Mauritania Niger Nigeria Rwanda Senegal Togo Uganda Zimbabwe	Burkina Faso Burundi Congo, Rep Mozambique Rwanda Sierra Leone Togo Zambia Zimbabwe
Poor	Mozambique Tanzania	The Gambia Madagascar Nigeria
Very poor	Sierra Leone Zambia	Ghana Kenya Malawi Tanzania

* see Appendix for definitions
Source: Adjustment in Africa and Table A3.

Monetary policy in Sub-Saharan Africa has focused primarily on maintaining low rates of inflation and adequate levels of real interest rates. These objectives were used as the basis for assessing the monetary policy stance in SSA. In this context, measures of the monetary policy stance, based on revenue from printing money (seigniorage), the rate of inflation, and the real interest rate (World Bank, 1994) were used in the analysis. Based on these indicators there has been an increase in the number of countries with 'good or adequate' policy (Table 3.2, Figure 3.2). There was also an increase in the number of countries with 'poor' and 'very poor' monetary policy by 1996-97. This may have been due to the inflationary pressure from significant currency devaluations. Although monetary policy appears to be more stable than the fiscal policies, instability remains fairly high. Ghana, Kenya and Malawi moved from a 'fair policy' stance in 1990-91 to a 'very poor' policy stance in 1996/97. The variability of policies among countries also remains high with nine countries having inflation rates over 20 percent. Seigniorage was above 2 percent in eight countries, while real interest rate varied between positive 10 percent in The Gambia and negative 10 percent in Nigeria (Table A3).

Box 3.1: Monetary Policy Effects on South African Agriculture.

Several studies have examined the effects of monetary policy on agriculture in South Africa, a country which has undertaken significant economic reform (Dushmanitch and Darroch, 1990 and Townsend, 1998). South Africa adopted a macroeconomic policy stance in the 1980s that was more market determined. This change was accompanied by agricultural sector reforms in the early 1990s. Dushmanitch and Darroch (1990), using a general equilibrium simultaneous equation model, found that an èxpansionary monetary policy caused real interest rates to fall, exchange rates to depreciate and the general price level to rise in the short run. They summarize their findings as "...depreciation of the exchange rate and higher domestic inflation raised input prices. Increased cost effects of higher input prices outweighed the reduced cost effects of lower real interest rates causing real field crop and horticultural supply to decrease.... The resulting decrease in agricultural supply caused commodity prices to rise, which lowered real demand for agricultural products. The net effect was a decline in real agricultural income for the crop, horticultural and livestock sectors" (Dushmanitch and Darroch, 1990). A possible explanation for this outcome is that marketing boards inhibited the pass-through of currency devaluation to higher domestic producer prices. Thus, macroeconomic policy effects worked primarily through the price of inputs.

At the aggregate level, these results are largely corroborated by Townsend (1998) using an impulse response function model (see Box 3.3 for a brief description of the model). The result of a positive shock to money supply is an asymmetric increase in the real price of agricultural inputs and outputs, with input prices increasing more than output prices. These effects have a corresponding negative impact on farm income (see below). This effect may be the result of relatively high levels of concentration in the input industry and the output price setting policies used by the marketing boards. The result of a positive shock on the real interest rate is a rise in real input prices as the input industries pass on the higher costs to farmers through higher prices. Similarly for output prices, marketing board costs may have been passed on to farmers as lower output prices. These changes have a significant effect on real net farm income (net of interest payments), which was exacerbated by the extremely high debt position held by most South African farmers at the time.

The Response of Agriculture to a Positive Shock in Monetary Policy Variables
(percentage deviation from the base line)

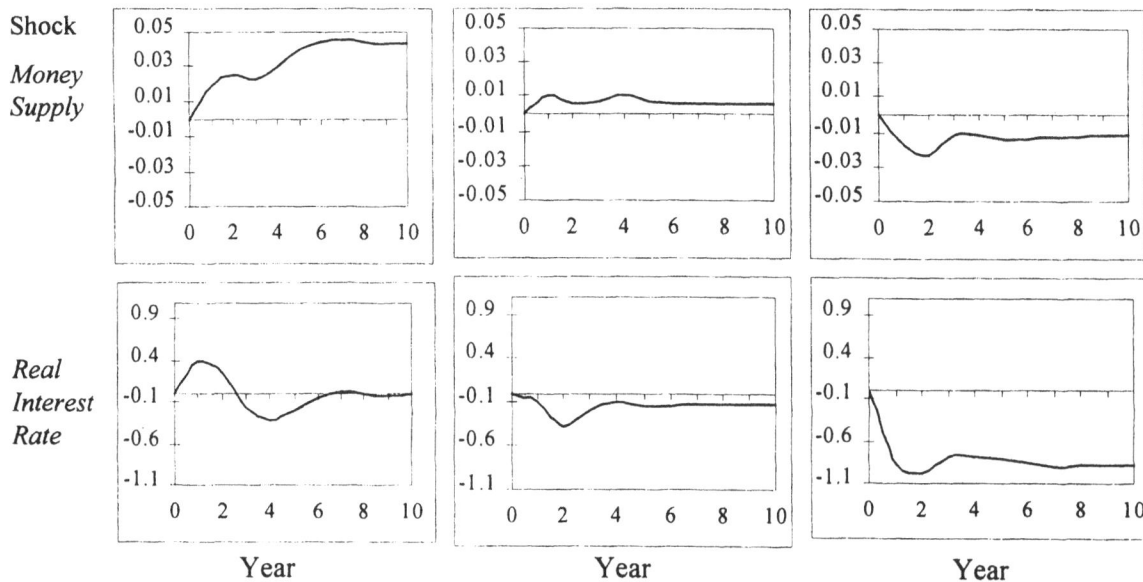

Response of:
Real Price of Agric. Inputs   Real Price of Agric. Outputs   Real Net Farm Income

EXCHANGE RATE POLICY

The structural adjustment programs adopted by most countries in Sub-Saharan Africa recognized that the tradable sector would be central to restoring economic growth. The corresponding policy prescriptions strongly emphasized the adoption of outward-oriented development strategies - especially export expansion. In Sub-Saharan Africa the policies pursued to achieve this objective included currency devaluations.

The overvaluation of domestic currencies was a common trait to almost all African economies during the 1970s and early 1980s. A combination of fixed nominal exchange rates and rapidly expanding aggregate demand resulted in large parallel market premiums. Excess demand for foreign exchange was controlled by imposing foreign exchange restrictions instead of devaluing the domestic currency. The result was a reduction in exports with an increasing scarcity of imports (Sahn *et al*, 1996). The aims of the subsequent reform programs were to promote exports and open up the economy to the benefits of international competition. Significant progress has been made in correcting overvalued exchange rates since the 1980s. Countries with flexible exchange rate regimes were able to devalue their currencies by reducing the large parallel market premiums, thus improving international competitiveness. The francophone countries in Africa also devalued the CFA franc by 50 percent.

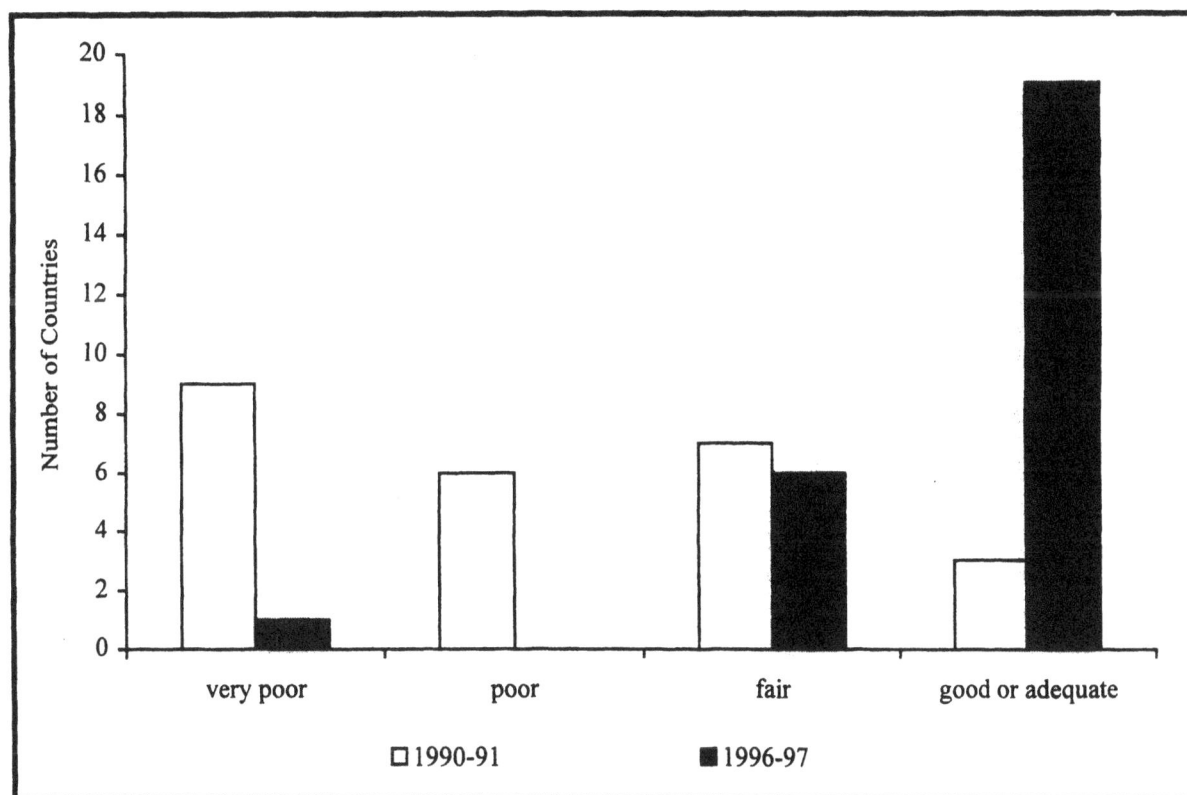

Figure 3.3: Comparison of the Exchange Rate Policy Stance in Sub-Saharan Africa between 1990-91 and 1996-97.
Source: World Bank (1994) with additional World Bank and IMF data for 1996-97.Table 3.3.

Table 3.3: Exchange Rate Policy Stance.

Policy Stance	1990-91	1995-97
Good or Adequate	Ghana Kenya Madagascar	Burkina Faso Cameroon Central African Republic Gabon Ghana Kenya Madagascar Malawi Mali Mauritania Mozambique Niger Rwanda Senegal Sierra Leone Tanzania Uganda Zambia Zimbabwe
Fair	Burundi The Gambia Malawi Nigeria Niger Uganda Zimbabwe	Benin Burundi Côte d'Ivoire Congo, Rep The Gambia Togo
Poor	Rwanda Benin Burkina Faso Central African Republic Gabon Mali Togo	
Very poor	Mauritania Mozambique Sierra Leone Tanzania Zambia Cameroon Congo Côte d'Ivoire Senegal	Nigeria

Source: Adjustment in Africa and Table A3.

The improvement of the exchange rate policy between 1990-91 and 1996-97 has been substantial (Figure 3.3, Table 3.3). Of the twenty-six countries analyzed, the number of countries with 'good or adequate' exchange rate policies increased from five to nineteen. Likewise, the number of countries with very poor exchange rate policies declined from nine to one. Nigeria was the only country with a 'very poor' exchange rate policy stance in 1996-97, caused largely by the collapse of the Nigerian financial system (Lewis and Stein, 1997).

Most of the countries analyzed had low parallel market exchange premiums (less than 10 percent) except for Burundi and Nigeria (Table A3). The countries with fixed exchange rates adjusted more slowly than countries with flexible exchange rates due to the difficulty of devaluing the domestic currency, which is pegged to the French franc. A comparison of francophone countries with a reference group of developing countries outside Africa, which exported primary products and did not have high parallel market premiums, suggested that these African countries needed to devalue their currency to remain competitive (World Bank, 1994). In 1994 the CFA was devalued by 50 percent against the French franc, a policy measure which has improved the export competitiveness of these countries.

Exchange rates are perceived to have the largest direct impact on agriculture than any other macroeconomic variable. Indeed, most studies examining the macroeconomic impacts on agriculture focus on this variable.

Studies have generally examined the impact on supply response (Jaeger, 1992) or focus directly on the impact of exchange rates on agricultural prices (Valdés, 1996, Box 3.2). Mamingi (1996) reviews the inclusion of an exchange rate variable in supply response models highlighting some of the estimation problems that have occurred (simultaneity bias and omitted variable bias).

Box 3.2: The Structure of Agricultural Prices.

Agricultural prices have been frequently decomposed into changes in the international price, changes in the nominal protection rate and changes in the real exchange rate (Valdés, 1996). Most of these studies focus on the evolution of the real producer price

$$p_{it} = \frac{P_{it}}{CPI_t} \qquad (1)$$

where P_{it} is the nominal price of agricultural good i at time t, measured in domestic currency and CPI_t is the consumer price index at time t. P_{it} can be further expressed as

$$P_{it} = P^*_{it} E_t (1+\gamma_{it})(1+t_{it}) \qquad (2)$$

where P^*_{it} is the corresponding border price the country faces (c.i.f. for importables and f.o.b. for exportables) measured in foreign currency (US dollars). E_t is the nominal exchange rate (measured in units of domestic currency per US$) at time t. γ_{it} is meant to be a 'mark-up' factor including transport costs and competitive profit margins to make the border price comparable with the domestic price. t_{it} is the residual after the mark-up and is meant to be the nominal protection rate.

Alternatively from equation (1) p_{it} can be expressed as

$$p_{it} = \left(\frac{P_{it}}{P^*_{it}E_t}\right)\left(\frac{P^*_{it}}{CPI^*_t}\right)\left(\frac{CPI^*_t E_t}{CPI_t}\right) = NPC_{it} * RWP_{it} * RER_{it} \qquad (3)$$

where CPI^*_t is the general level of the foreign prices at time t (US CPI). NPC is the nominal protection coefficient, RWP is the real world price and RER is the real exchange rate. Using equation (2) and (3) this can be expressed as

$$p_{it} = (1+\gamma_{it})(1+t_{it})P^*_{it} RER_t \qquad (4)$$

where RER denotes the real exchange rate, defined as the ratio of international domestic prices. Equation (4) can be rearranged as

$$\frac{p_{it}}{P^*_{it} RER_t} = (1+\gamma_{it})(1+t_{it}) \qquad (5)$$

The right hand side of this expression corresponds to a hypothetical transport cost and competitive profit margin, explicit export and import tariffs and implicit import and export tariffs. This methodology has been commonly used to decompose real domestic prices into changes in external and domestic factors. The external factors are reflected in the world price while the domestic factors are reflected in the change in the real effective exchange rate and the nominal protection coefficient.

Box 3.3: Examples of Exchange Rate Effects on Agriculture in Sub-Saharan Africa.

Inelastic supply elasticities

The exchange rate is central to the macroeconomic effects on agriculture and several authors have estimated its effect on agricultural production in SSA. One often cited study by Jaeger (1992), uses data from 1970-87 to estimates supply functions for several crops. Real exchange rates were included as an explanatory variable and its elasticity was estimated as -0.10 for all countries and all crops, suggesting that a 10 percent real exchange rate depreciation results in a 1 percent increase in output. This elasticity was calculated as -0.35 for cocoa, -0.68 for cotton and -0.25 for all tree crops.

Past policy distortions and exchange rate pass-through to agricultural producer prices in South Africa

Like many African countries the agricultural output markets in South Africa were dominated by marketing boards. Producer prices were controlled and farmers were insulated from the external market. Townsend (1998) analyzed the effects of currency devaluations on the agricultural sector in view of these interventions.

A VAR model was used in the analysis, which included variables of money supply, real interest rates, the real exchange rate, real agricultural input and output prices, real net farm income and agricultural exports. Data from 1947-1995 were used and the statistical properties of all the time series variables included in the analysis were examined (all variables were integrated of order one). Cointegration tests also suggested that there was a valid long-run relationship between the macroeconomic variables and the agricultural variables included in the model (this is the same model used for the results presented in Box 3.1 – for brevity, only the shock of the exchange rate on real agricultural input and output prices and real net farm income are reported below).

Using the VAR model an impulse response function was applied to examine effects of an exchange rate shock on the real price of agricultural inputs, outputs and real net farm income. A positive shock on the exchange rate (a devaluation of the Rand) had a significant impact on increasing the real input price while its effect on the output price tended to be muted. Thus the corresponding effect on real net farm incomes was through the input price rather than the output price, an effect exacerbated by concentration in the agricultural input industries. This clearly shows that the exchange rate shocks were not passed though to output prices and export producers were faced with distorted incentives. The lack of pass-through of the currency devaluation to output prices was largely due to the fact that marketing boards insulated producers from external demand.

The response of agricultural prices and incomes to devaluations of the Rand
(*percentage deviation from the base line*)

Shock	Response of: *Real Price of Agric.* *Input*	*Real Price of Agric.* *Output*	*Real Net Farm Income*
Exchange Rate Depreciation			

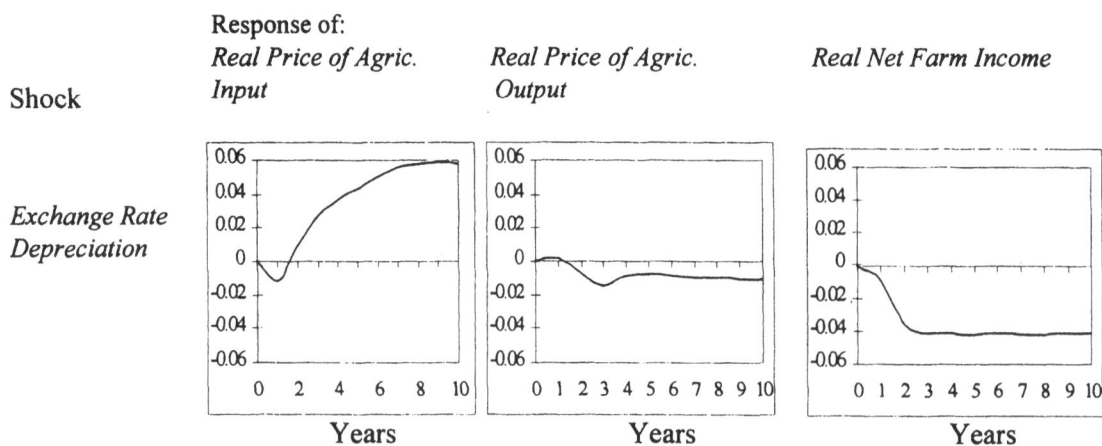

Years   Years   Years

OVERALL MACROECONOMIC POLICY STANCE

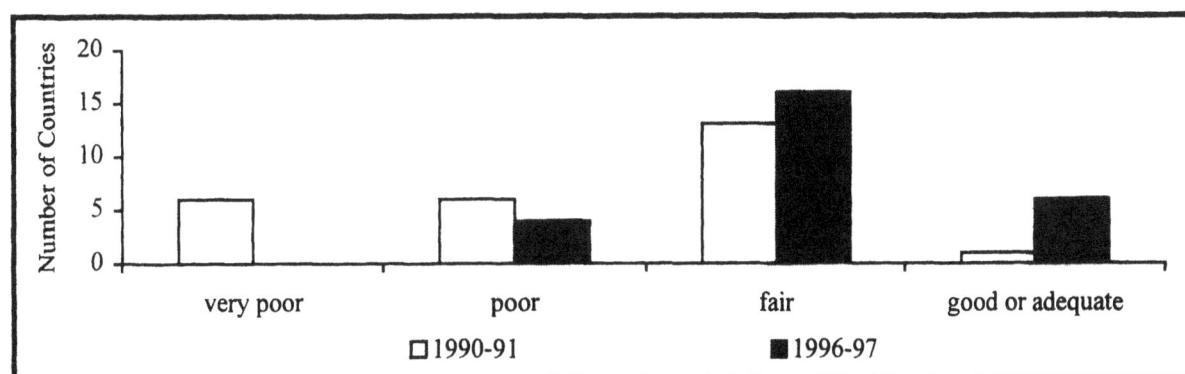

Figure 3.4: Comparisons of the Overall Macroeconomic Policy Stance between 1990-91 and 1996-97.
Source: Table 3.4.

Table 3.4. Macroeconomic Policy Stance.

Policy Stance	1990-91	1996-97
Adequate	Ghana	Gabon Mauritania Cameroon Senegal Niger Benin
Fair	Burundi The Gambia Madagascar Malawi Burkina Faso Kenya Gabon Mauritania Nigeria Senegal Togo Mali Uganda	Central African Republic Mali Uganda Burkina Faso Côte d'Ivoire Congo, Rep Zambia Kenya Madagascar Mozambique Rwanda Tanzania Sierra Leone Zimbabwe Togo Malawi
Poor	Central African Republic Niger Benin Rwanda Tanzania Zimbabwe	Burundi The Gambia Nigeria Ghana
Very poor	Côte d'Ivoire Cameroon Congo, Rep Mozambique Sierra Leone Zambia	

Source: Adjustment in Africa and Table A3.

The fiscal, monetary and exchange rate policy stances were combined to create a measure of the overall macroeconomic policy stance. Although this index is imperfect, it does provide a summary of macroeconomic policies (World Bank, 1994). Since 1990-91, there has been a significant improvement in the overall macroeconomic policy stance of countries in Sub-Saharan Africa (Table 3.4). Six countries had an 'adequate' policy stance in 1996-97, which represents an improvement from one in 1990-91 (Table 3.4). There was no country with 'very poor' macroeconomic policies and only four countries (Burundi, The Gambia, Nigeria and Ghana) showed 'poor' policies. Although there has been a general improvement, several countries experienced a high level of instability in their policies (Ghana, Nigeria). The CFA countries have significantly improved their overall policy stance, which can be largely attributed to the devaluation of the CFA franc in 1994. Of the twenty-six countries analyzed, nineteen made advances since 1990-91, while the macroeconomic policies deteriorated in four countries (Ghana, Burundi, Nigeria, The Gambia) (Table A3). As the *Adjustment in Africa* study shows, there was significant improvements in macroeconomic policy during the 1980s and as seen here, the improvement has continued in the 1990s.

As mentioned earlier, trade policy was not explicitly included in the overall macroeconomic policy score as it will be examined and included in the policy scores developed for export crops, food crops and fertilizer in later sections of the study. However, several key issues need to be raised in this macroeconomic policy section.

TRADE POLICY

Africa has made significant progress in reducing overvalued exchange rates while improving the overall macroeconomic policy stance. Foreign exchange rationing has been removed and balance of payments have become more sustainable. These improved policies have allowed significant import restrictions to be removed. Despite these improvements, many countries continue to inhibit trade via high import tariffs, a policy consistent with past inward-looking industrialization strategies where domestic industries were protected from external competition.

Sub-Saharan Africa's trade barriers (non-tariff barrier [NTB] coverage ratio and tariff levels) are several times higher than fast growing exporters (Figure 3.5). Tariffs on all imports average almost 27 percent for SSA, while, for the fast growing exporters and high income non-OECD countries the corresponding tariff levels were 9 and 3 percent respectively (Ng and Yeats, 1996).

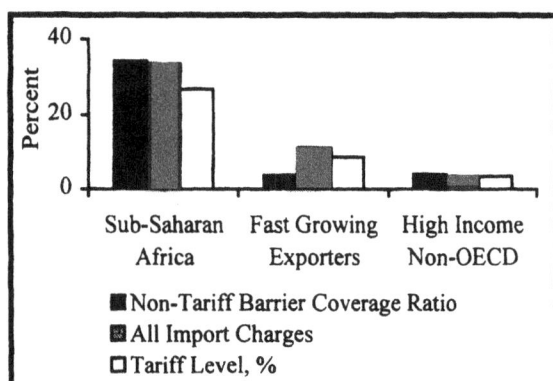

Figure 3.5: African Trade Barriers.
Source: Ng and Yeats (1996).

The tariff levels on imports used as inputs in agricultural production are also much higher than in the fast growing exporters (Figure 3.6). These tariffs represent an additional direct cost to exporters, who use these imports as intermediate inputs in production. The high costs of importing machinery and spare parts raise transport costs, which are extremely high in African countries. These high import tariffs place domestic producers at a cost disadvantage relative to competing countries.

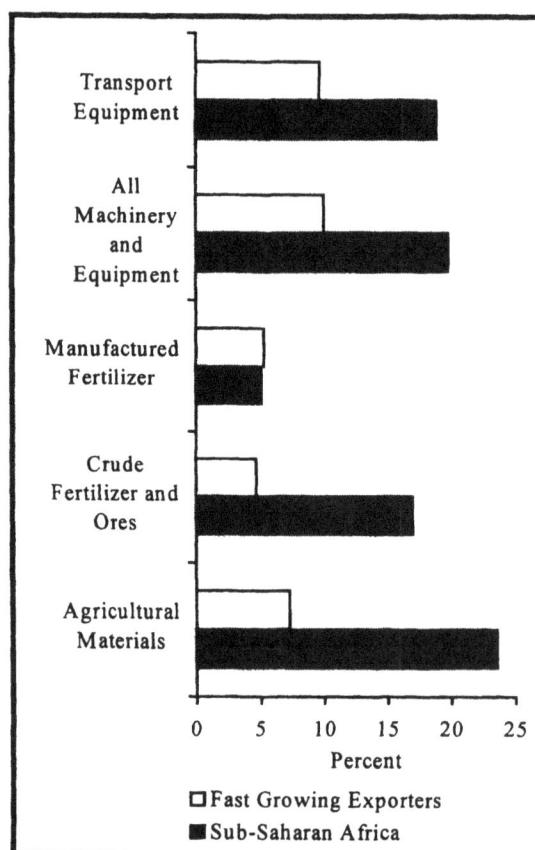

Figure 3.6: Tariffs on Factors of Agricultural Production.
Source: Ng and Yeats (1996).

The fast-growing exporters included in the comparison were Bahrain, China, Hong Kong, Indonesia, Jordan, Kuwait, Malaysia, Mexico, Qatar, Papua New Guinea, Republic of Korea, Saudi Arabia, Singapore, Taiwan and Thailand (Ng and Yeats, 1996).

Pledges for SSA countries to reduce these high tariff have been fairly weak as shown by the Uruguay Round commitments. Most tariffs are bound between 50–100 percent (Figure 3.7).

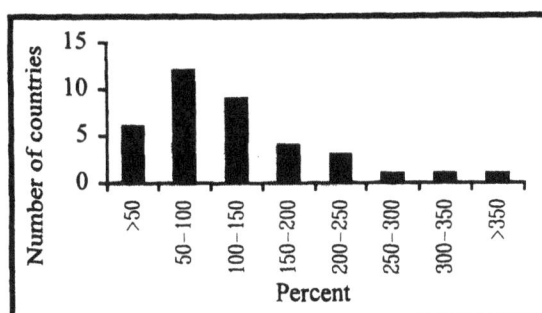

Figure 3.7: Sub-Saharan Africa Tariff Bounds from the Uruguay Round.
Source: Ingco and Townsend (1998).

INSTITUTIONAL FRAMEWORK AND THE CREDIBILITY OF RULES

Apart from monetary, fiscal, exchange and trade policies there is an expanding literature which emphasizes the reliability of policies (stability and uncertainties surrounding their implementation) as a factor affecting growth (Barro, 1991, Cukierman *et al*, 1996 and Brunetti *et al*, 1998). An environment characterized by unclear property rights, constant policy changes and policy reversals, uncertain contract enforcement and high corruption translates into lower investment and growth (Brunetti, *et al*, 1998). In these environments, the private sector is usually reluctant to commit resources especially in projects that require large sunk costs.

The 1997 World Development Report (WDR) constructs a credibility index based on a survey of local entrepreneurs in sixty-nine countries. The survey contained questions on the predictability of laws and policies, the subjective evaluation of political instability, the security of property and persons, the reliability of judicial enforcement and uncertainty stemming from corruption and bureaucratic discretion (see Brunetti et al, 1998 for its construction). The credibility index is thus based on investor's perceptions.

Figure 3.8 shows regional averages of the credibility indicator, which ranges in value from one (no credibility) to six (perfect credibility). Results of the survey show that the Commonwealth of Independent States has the lowest credibility index followed by Sub-Saharan Africa. Twenty-two African countries were included in the survey. Their credibility rankings are presented in Figure 3.9. The levels of credibility significantly affects growth and investment in particular countries. Regression results from the WDR, 1997, show that the credibility variable had a highly significant and positive effect on growth in GDP per capita and the investment/GDP ratio. Thus, while correct macroeconomic policies are a necessary condition to generate sustained economic growth, credibility (or the reliability of the countries institutional framework) is also critical, particularly in African countries. Thus, approaches aimed at improving traditional macroeconomic policy instruments should also include strategies to improving credibility.

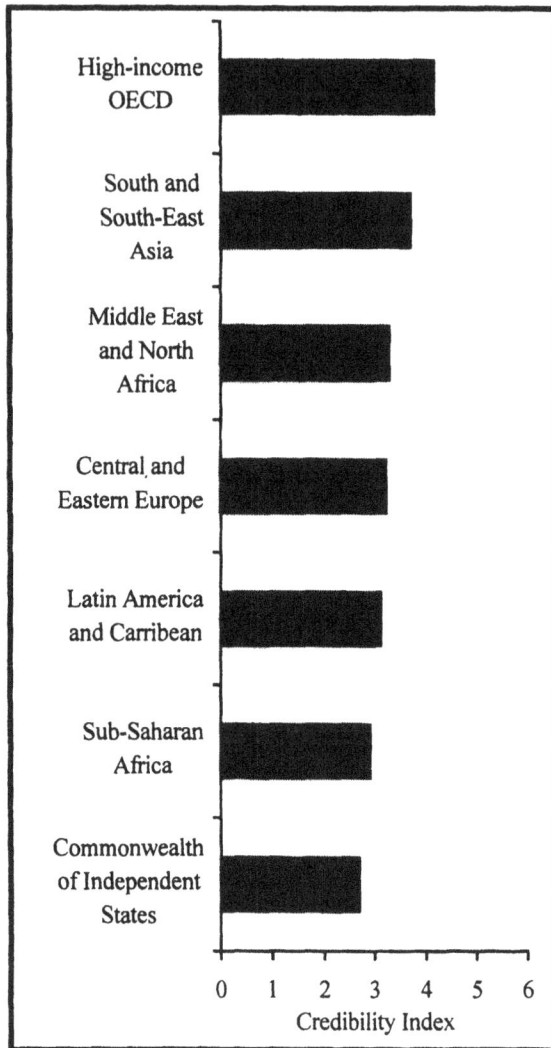

Figure 3.8: Regional Credibility Ratings, 1996.
Source: Brunetti *et al*, 1998.

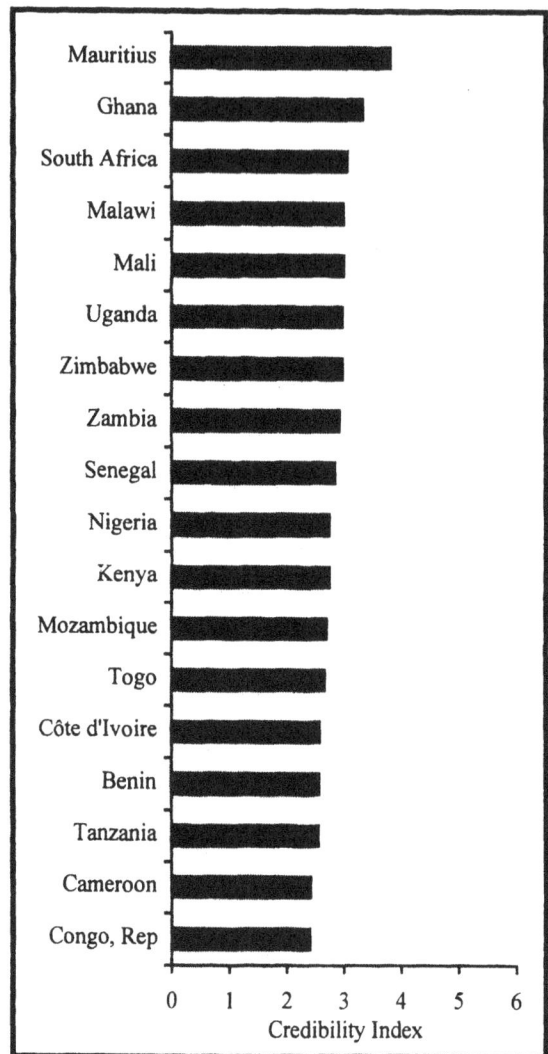

Figure 3.9: Sub-Saharan Africa Country Credibility Ratings, 1996.
Source Brunetti *et al*, 1998.

SUMMARY AND CONCLUSIONS

The linkage between macroeconomic policy and agricultural performance has been increasingly recognized in the 1990s. Overvalued exchange rates and industrial protection excessively taxed agriculture (Schiff and Valdés, 1992) distorting invectives and inhibiting growth. Agricultural policy has also impacted on the macroeconomy with heavy state intervention through marketing boards and price subsidies placed unmanageable pressure on fiscal balances. The macroeconomic environment has improved significantly in the 1980s which has continued through the 1990s although, for some countries, instability remains high.

- Fiscal policy stance scores showed that in 1996/97 eight countries had a 'good or adequate' stance which is an improvements from five countries in 1990-91. Only three countries (Ghana, The

Gambia and Zimbabwe) had a budget deficit of over 7 percent of GDP, which is a reduction from eight countries in 1990-91. A high level of instability remains in several countries.

- The monetary policy stance has also improved since 1990-91 when sixteen countries were classified as having 'fair' policies and six countries were classified as having 'good or adequate policies'. The corresponding numbers for 1996-97 were nine and ten. The number of countries with 'very poor' policies increased from two to four. This slight decline could be due to inflationary pressures from currency devaluation.

- The exchange rate policy stance has undergone an impressive improvement. In 1996-97 nineteen countries were classified as having a 'good or adequate' exchange rate regime. The corresponding number in 1990-91 was five. Nigeria was the only country in the sample with a 'very-poor' exchange rate policy stance due to the large parallel market exchange rate premiums in 1996/97.

- The overall macroeconomic policy stance has improved significantly. In 1996-97 eight countries were classified as having an adequate macroeconomic policy stance. This is an improvement from one in 1990-91. No countries were classified as having a 'very poor' policy stance which is an improvement from six countries in 1990-91. Again, in several countries, a fairly high level of instability remains.

- Import tariffs on inputs used in agricultural production (transport equipment, manufactured and crude fertilizer and agricultural materials) remain several times higher in SSA than in the fast growing exporting countries.

- The reliability of the institutional framework (credibility) in Sub-Saharan Africa in among the lowest in the world (second only to the CIS), varying significantly across the continent. As a result, growth has been stifled.

While macroeconomic policies have continued to improve, there needs to be a continued focus on maintaining macroeconomic stability in many countries. There also needs to be an increased focus on improving the credibility of rules and the reliability of the institutional framework. In many African countries, low credibility has inhibited private sector investment and stifled growth. If the private sector is to be encouraged to enter the market and take over some of the activities in agriculture previously performed by the state, then this fundamental issue should be high on the policy agenda. The improvement in macroeconomic policies also needs to be passed-through to farmers for them to fully benefit from these reforms. These issues will be examined more closely in the next section.

4. EXPORT CROP POLICIES, PRICES AND MARKETS

Since the early 1980s, the widespread adoption of structural adjustment programs by African countries resulted in significant policy reforms in the export crop sector. Common elements of these reforms included removing distortions that prevented markets from functioning efficiently. The emphasis was placed on: i) eliminating price controls; ii) developing competitive local markets; iii) reducing state intervention in international trade to enhance integration into world markets; and iv) improving aspects of the regulatory system and privatizing inefficient public enterprises (Meerman, 1996). As discussed in the first chapter, the theory behind these programs was to improve market efficiency and the output-input price ratio in order to induce an increase in agricultural output.

PRICE RESPONSE TO REFORMS

Evidence suggests that there has been a significant price response to these reforms. In the 1980s, however, this positive response was eroded by the decline in real world commodity prices making it difficult for African governments to improve producer price incentives. The *Adjustment in Africa Report* (World Bank, 1994) shows that ten of the twenty-seven countries analyzed managed to increase real producer prices from 1981-83 to 1989-91 (Figure 4.1), the achievement of which was attributed to the adoption by these countries of better policies. Some of these ten countries gave producers a larger share of border prices by lowering export taxes, raising administered producer prices, reducing marketing costs, or liberalizing marketing (World Bank, 1994).

In several countries that experienced declining real producer prices (Burundi, the Central African Republic, Republic of Congo, Gabon, Kenya, Malawi, Sierra Leone, Uganda and Zimbabwe), governments had taken measures to help their farmers. However, these were not sufficient to offset the decline in world prices. In other countries, (Cameroon, Chad, Côte d'Ivoire, The Gambia, Guinea-Bissau, Rwanda, Senegal and Zambia), the combined effect of macroeconomic and sectoral policies worked against agriculture and compounded the decline in export prices. About two thirds of these countries reduced the overall tax burden on agriculture. However, these countries either reduced the explicit taxation or the implicit taxation, but not both (World Bank, 1994).

Figure 4.2 provides a comparison, showing the change in the real producer price of export crops in the 1990s. There has been a large improvement in the number of countries experiencing real domestic producer price increases. This favorable trend can largely be attributed to both the increase in world commodity prices (Table 2.5) and the improvement of both agricultural and macroeconomic policies. Some of these policy improvements will be further discussed in this chapter. Fifteen of the nineteen countries included in the comparison experienced an increase in the real domestic producer price of agricultural export crops (Benin, Burkina Faso, Cameroon, Côte d'Ivoire, Ghana, Nigeria, Madagascar, Malawi, Mali, Mozambique, Senegal, Tanzania, Togo, Uganda, Zimbabwe,). Only four countries (Burundi, Chad, The Gambia, Kenya) experienced price declines.

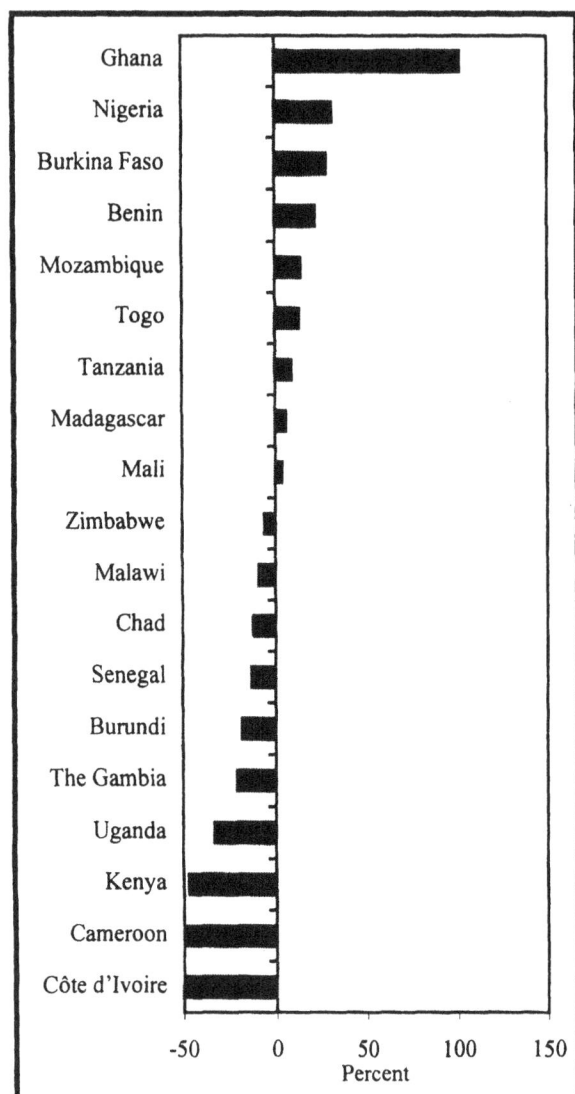

Figure 4.1: Change in the Real Producer Price of Agricultural Exports, 1981-83 to 1989-91.
Source: World Bank (1994).

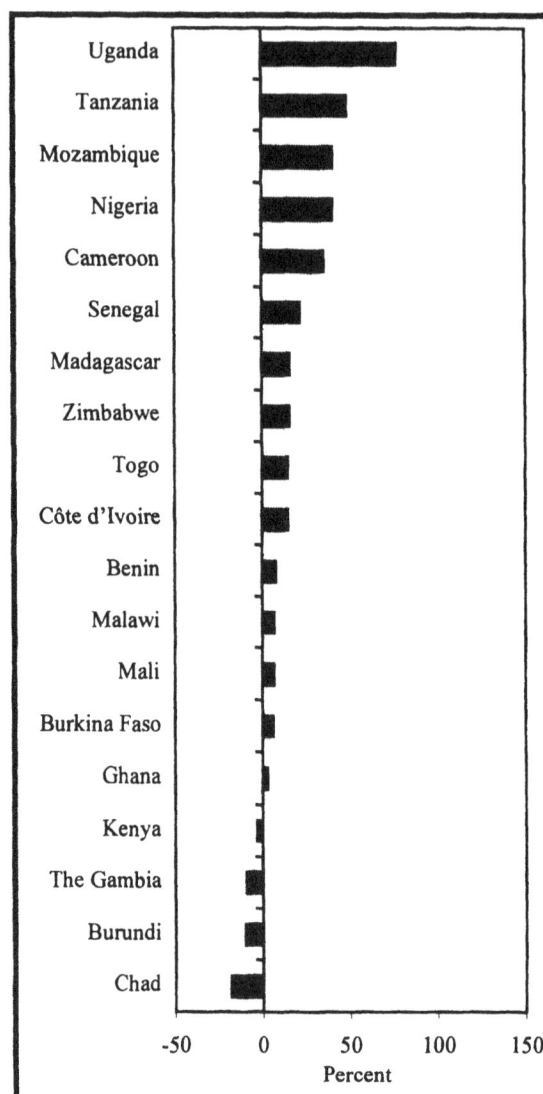

Figure 4.2: Change in the Real Producer Price of Agricultural Exports, 1989-91 to 1995-97.
Source: IMF and World Bank data.

Data for each country are based on the percentage change in the real producer price of export crops. The aggregate price changes were derived as a weighted average of the major export crops. The Adjustment in Africa Report (World Bank, 1994) included other countries in its analysis, but reliable price data for these other countries were not readily available for 1990-91 to 1995-97.

If these prices are examined over the 1981-97 period, twelve out of the nineteen countries experienced real producer price increases. The countries in which prices declined over this longer period were Burundi, Cameroon, Chad, Côte d'Ivoire, Malawi, The Gambia and Kenya. Nine countries experienced an improvement in the real producer price of exports in both time periods [during the 1980s and 1990s] (Ghana, Nigeria, Tanzania, Mozambique, Burkina Faso, Benin, Togo, Madagascar, Mali); three countries had an improvement in the 1990s, which offset the price declines they experienced in the 1980s (Uganda, Senegal, Zimbabwe). Cameroon, Côte d'Ivoire and Malawi experienced an increase in domestic prices between 1989-91 and 1995-97, but these increases were not large enough to offset the price declines in the 1980s. The Gambia, Burundi, Chad and Kenya experienced price declines in both periods (Table 4.1).

Table 4.1: Real Export Producer Price Performance in Sub-Saharan Africa.

'Advancing' (9)		'Caught up' (3)	'Catching up' (3)	'Lagging' (4)
Ghana	Benin	Uganda	Cameroon	The Gambia
Nigeria	Togo	Senegal	Côte d'Ivoire	Kenya
Tanzania	Madagascar	Zimbabwe	Malawi	Burundi
Mozambique	Mali			Chad
Burkina Faso				

Source: Data from Figure 4.1 and 4.2.

'Advancing' - countries that experienced and increase in the real export producer price for both time periods (1981-83 to 1989-91 and 1989-91 to 1995-97), 'Caught up' - increases in the real export producer prices between 1989-91 and 1995-97 exceeded the decline between 1981-83 and 1989-91, resulting in a net improvement. 'Catching up' - means that the increase in the real export producer prices between 1990-91 to 1995-97 have not yet exceeded the decline from 1981-83 to 1989-91, resulting in a net decline, 'Lagging' countries mean that these countries experienced a decline in the real producer price for both periods.

The structure of these price incentives can be examined by decomposing the change in real producer prices into three elements: the change in the real world price; the change in the real exchange rate and the change in the nominal protection coefficient (NPC) which, in this case, is simply the change in the producer's share of the world price (Box 3.2). The nominal protection coefficient is the ratio the farmer gate price to the domestic currency border price adjusted for transportation and marketing costs. Thus, the NPC should reflect the extent of direct taxation or subsidy (Box 3.2). NPCs have typically been poorly calculated with many studies simply using the producer's share of the border price. The common assumption is that overtime transportation and marketing costs remain relatively fixed and thus the movements in the producer's share of the border price is due to the extent of taxation or subsidy. In Africa, data on transportation and marketing costs overtime are particularly difficult to obtain, and the quality of these data, when available, are very poor.

The share of each component in equation 3 in Box 3.2 can be derived using a log transformation. The change in the real producer price resulting from a change in these components can be computed as $\hat{p}_t = R\hat{W}P_t + R\hat{E}R_t + N\hat{P}C_t$. The hats represent percentage changes and are calculated as differences of the natural logarithms of the variables. *RWP* is the real world price which is usually attributable to external factors, *RER* is the real effective exchange rate, which is affected by government macroeconomic policy and by external shocks while *NPC* is determined by domestic price, marketing and trade policy.

This decomposition for the period 1981-83 to 1989-91 is shown in Figure A3 in the appendix and for the period 1990-91 to 1995-97 in Figure A4 in the appendix (see Table A8). During the 1980s African countries were faced with significant downward trends in real world prices (RWP). The decline had a significant negative impact on producer prices (Figure 4.3) but despite these adverse effects, nine countries managed to increase price incentives for export crops. This was largely achieved through significant depreciation of the real exchange rate (particularly in Ghana, Nigeria, Tanzania and Uganda) and an improvement in agricultural marketing, price and trade policy. Macroeconomic policy changes had a large positive effect on prices but at the aggregate, the producer's share of the border price (NPC)

47

barely changed which lets us conclude that in many countries sectoral policy did little to improve farm level export crop prices.

During the 1990s the situation changed somewhat, real world prices became more favorable and improvements in sectoral policy contributed more to price increases. Between 1989-91 and 1995-97, macroeconomic policies continued to improve but these changes were less dramatic than in the 1980s as the space for further policy improvement had been reduced (Figure 4.4). Five countries experienced an appreciation of their real exchange rate (Nigeria, Madagascar, Kenya, Burundi and Tanzania [Figure A4]) which placed downward pressure of domestic export prices. Of these countries, the improved agricultural policies in Madagascar, Nigeria and Tanzania were sufficient enough to offset these negative price pressures. Nine countries improved their agricultural policies with significant improvements in Uganda, Tanzania, Madagascar, Mozambique, Nigeria, Malawi, Zimbabwe and The Gambia. In Benin, Togo, Burkina Faso, Burundi and Chad the producer's share of the border price declined. This trend suggests that in these countries, the more favorable macroeconomic policies have not been fully transmitted to producers as higher prices. Indeed in some countries agricultural policies have eroded the price benefits from more favorable world prices and have inhibited the pass-through of exchange-rate depreciations to producer prices.

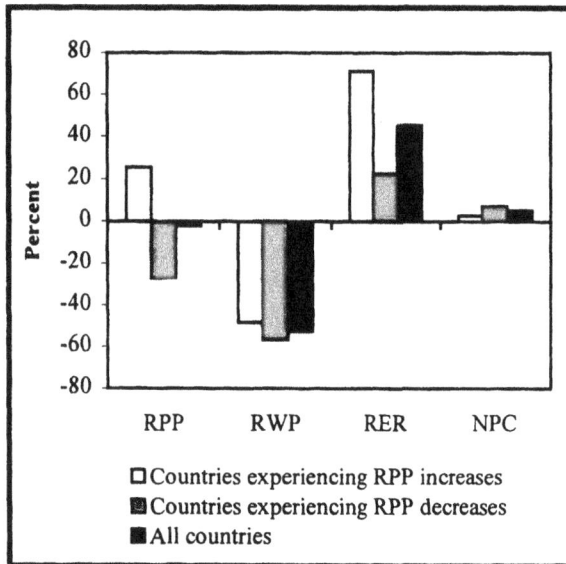

Figure 4.3: The Structure of Price Incentives, 1981-83 to 1989-91.
Source: Table A8.

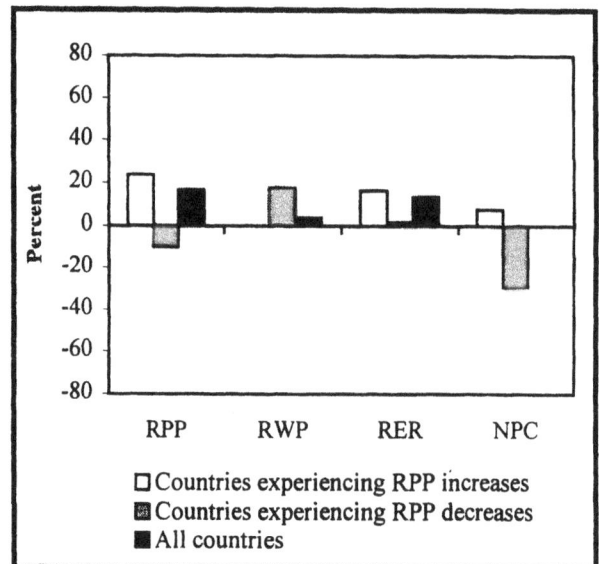

Figure 4.4: The Structure of Price Incentives, 1989-91 to 1995-97.
Source: Table A8.

RPP = real producer price for export crops, RWP = the real world price, RER = the real exchange rate and NPC = nominal protection coefficient (in this case, the producer's share of the border price).

SUPPLY RESPONSE TO PRICES

The perceptions behind the reform programs were that farmers were price responsive and that the removal of distortions affecting producer price incentives would induce higher production levels. Studies on the supply response of African farmers have expanded tremendously over the past several years (Binswanger, 1987, 1992; Jaeger, 1992; Maminigi, 1998 provide some examples). The general consensus is that African farmers are price responsive and that policy reforms have resulted in that in the short run, price elasticities range between 0.1 and 0.8 (Table A10) for individual crops. However, the responsiveness of aggregate agricultural production to price changes is still very low in Africa (Bond, 1983; Chhibber, 1989; Binswanger, 1989). Bond (1983) estimated the average price elasticity of total agricultural production for nine African countries to be only 0.18 in the short run and 0.21 in the long run. Chhibber's (1989) review of empirical studies concluded that in low-income developing countries with poor infrastructure, the aggregate long-run price elasticity of supply was within the 0.3-0.5 range. Binswanger (1989) estimates a short run elasticity of 0.06.

While 'getting prices right' is a necessary condition it is by no means sufficient to induce supply response. In the absence of other supporting non-price policies, it is alone seldom sufficient to bring about the desired changes in the agricultural sector (as discussed in chapter 1). This is not to deny evidence that African farmers, both small and large-scale, are price-responsive.

In a recent assessment of supply response, Meerman, (1996, pg. 2), summarizes his findings as "...Supply response was found to be...

- *Symmetrical* – The removal of heavy agricultural protection leads to contraction of production with an accelerated movement of resources away from the previously protected crops [ie: maize production in South Africa] to more profitable crops.
- *Synergistic* – The level of supply response from economic reform depends on the degree to which the agricultural economy is developed. Adequate rural infrastructure (irrigation, roads and transport, power and telecommunications), input availability, research, credit and farmer education and health are conducive to agricultural development. Where these are seriously deficient, even getting the prices right in an ideal enabling environment will not suffice to get agriculture moving. [Price responsiveness also depends on the agro-ecological zone and the labor economy (household labor capacity, gender composition and labor organization)].
- *Dependent on the credibility of reforms* – The private sector does not invest if the continuity of the reforms is in doubt. Reform programs have frequently been reversed or halted. This has allowed public enterprises to dominate the marketing system and control the export of important crops. Government policy has frequently been unpredictable. Thus establishing the credibility of policy measures is at least as important as choosing the efficient policy solution".

With these factors in mind, there is some evidence of improved export crop supply. Since 1990, agricultural exports from SSA have increased by about 30 percent. However, large fluctuations have occurred in export volumes (Figure 4.5) due to the effects of the weather with drought years in 1983, 1991 and 1995. These export trends over time show a very similar pattern to overall agricultural growth (Figure 1.1 in Chapter 1) suggesting that overall growth is strongly influenced by the performance of agricultural exports. Indeed, it is very difficult to improve agricultural growth if producers are confined to local or domestic markets.

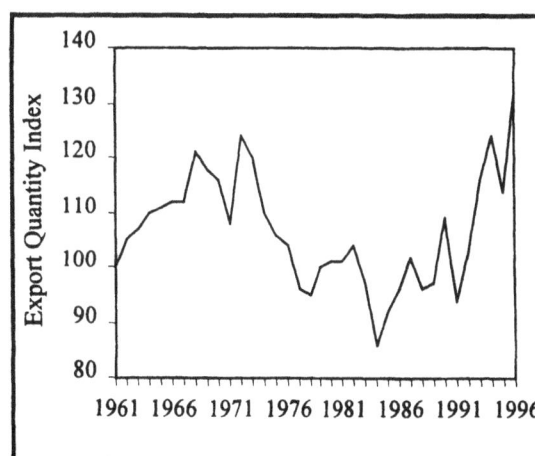

Figure 4.5: Sub-Saharan African Agricultural Export Volume.
Source: FAO data.

As seen from the analysis so far, more favorable world prices and continuing improvements in macroeconomic policies have contributed significantly to improving producer price incentives. Although the space for further improvements in these macroeconomic policies has been reduced, there appears to be room for improving domestic agricultural and broader rural development policies. During the 1980s the contribution of these domestic policies (factors) was limited (Figure 4.3 and 4.4). This, however, improved in the 1990s but the contrasts of these factors among African countries remains large. Improving the producer's share of the border price will require not only an improvement in policy (agricultural, trade and regulatory) but will also require a reduction in transportation and transaction costs to improve market efficiency and price transmissions. This again highlights the complementarities between price and non-price factors. These issues will be examined this the next section.

ENHANCING MARKET EFFICIENCY AND PRICE TRANSMISSIONS

A key component of structural adjustment programs was to improve market efficiency through policy (trade, agricultural and regulatory) reform. The objective was to reduce government interventions that distorted prices and tied up markets (World Bank, 1994). The theoretical notion of efficient markets is that: i) if there are enough markets; ii) if all consumers and producers behave competitively; and iii) if an equilibrium exists, then the allocation of resources in that equilibrium will be Pareto[17] optimal (Ledyard, 1987). The proposal is that if input and output markets are complete, so that no transactions are missed, and if there are so many buyers and sellers that none can alone influence prices, then the market outcome will be efficient. In this static framework, efficiency means that all three conditions will be met simultaneously (Ward, Deren and d'Silva, 1990). These are: i) resources will be fully employed; ii) they will be correctly allocated to productive enterprises so that each output will be efficiently produced; and iii) the correct combination of outputs will be produced, meaning the combination that will maximize the welfare of consumers, given the initial distribution of resource ownership, and hence incomes.

In Africa, these conditions rarely exist with missing markets, imperfect information and high transaction costs. While difficult to measure, it is important to provide a characterization of the market environment in African countries. Proxy indicators could provide some useful information. One possible proxy is the exchange rate pass-through to producer prices with cross-country differences in this variable providing some indication of the difference in market structure, product characteristics, competition and trade barriers (Menon, 1995).

EXCHANGE RATE PASS-THROUGH TO PRODUCER PRICES

The preceding analysis on prices clearly showed that between 1990 and 1997 many farmers have experienced an increase in their real price of agricultural exports. Part of these price increases were shown to be the result of domestic currency devaluations and in some countries improved domestic agricultural policies. This section will analyze these domestic policy issues more closely focusing on the factors that have inhibited the exchange rate pass-through to producer prices. In some countries, the pass-through of currency devaluations to producers has been limited (Figure 4.6 Table 4.2) with a contrasting success of pass-through between crops and countries within the region.

---

[17] Pareto optimality means that social welfare is maximized in the limited sense that nobody can be made better off without making someone else worse off.

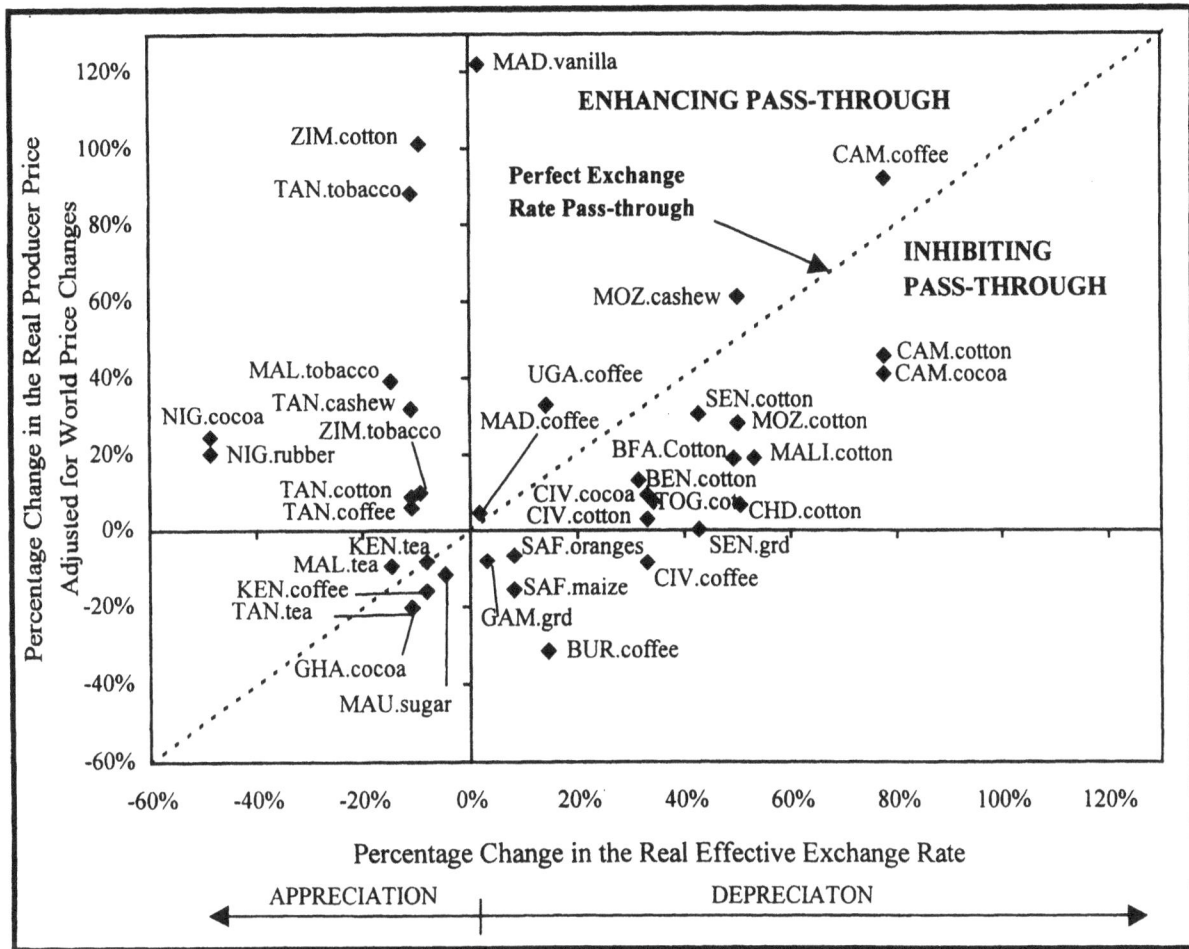

Figure 4.6: Pass-Through of Real Effective Exchange Rate Depreciations to Real Agricultural Export Prices, 1990-1995/97.
Source: IMF and World Bank data. See Table 4.2 for full country names.

Table 4.2 summarizes these findings. In some cases exchange rate pass-through has been enhanced, while in other cases it has been inhibited. Exchange rate pass-through refers to the degree to which exchange rate changes are reflected in the domestic currency prices of traded goods. The literature on exchange rate pass-through has expanded significantly over the last two decades but much of the focus has been on developed countries (see Menon, 1995 for a summary). Many of these earlier studies have focused on the impact of currency devaluation on domestic import prices.

Figure 4.6 and Table 4.2, shows that exchange rate pass-through has been enhanced for many crops in several countries. Vanilla farmers in Madagascar received the greatest pass-through of exchange rates (120 percent), this is due to the significant increase in the farmer's share of the f.o.b. price, which was only 8 percent in 1990. Similarly, cotton farmers in Zimbabwe, tobacco and cashew farmers in Tanzania, cocoa and rubber farmers in Nigeria and tobacco farmers in Malawi experienced greatly enhancing pass-through. At the other extreme, the exchange rate pass-through was greatly inhibited for cotton farmers in Chad, coffee farmers in Burundi and Côte d'Ivoire and groundnut farmers in Senegal. As will become evident, the extent of exchange rate pass-through is highly correlated with market interventions.

Table 4.2: Exchange Rate Pass-Through to Producer Prices, 1990 to 1995-97.

Enhancing Pass-through			Inhibiting Pass-through	
Greatly enhancing			*Greatly inhibiting*	
Madagascar - vanilla	120%		Burundi – coffee	-46%
Zimbabwe - cotton	111%		Chad – cotton	-44%
Tanzania - tobacco	99%		Senegal – groundnuts	-42%
Nigeria - cocoa	73%		Côte d'Ivoire – coffee	-42%
Nigeria - rubber	69%			
Malawi - tobacco	54%		*Inhibiting*	
Tanzania - cashew	43%		Cameroon – cocoa	-37%
			Mali - cotton	-34%
Enhancing			Cameroon – cotton	-32%
Tanzania - cotton	20%		Côte d'Ivoire – cocoa	-30%
Zimbabwe - tobacco	19%		Burkina Faso – coffee	-30%
Uganda - coffee	19%		Togo - cotton	-27%
Tanzania - coffee	17%		South Africa – maize	-24%
Cameroon - coffee	14%		Côte d'Ivoire – cotton	-24%
Mozambique - cashew	11%		Mozambique - cotton	-22%
			Benin – cotton	-18%
Near congruence			South Africa – oranges	-15%
Malawi - tea	5%		Senegal – cotton	-12%
Madagascar - coffee	3%		The Gambia - grd	-11%
Kenya - tea	1%			
			Near congruence	
			Ghana – cocoa	-9%
			Tanzania – tea	-9%
			Kenya – coffee	-8%
			Mauritius – sugar	-7%

The percentages represent the difference between the percentage change in the real producer price (adjusted for the world price) and the percentage change in the real effective exchange rate.
Source: Calculated from IMF and World Bank data, Figure 4.6.

A general observation from Table 4.2 is that where producers have experienced (greatly) enhancing pass-through, domestic markets have opened up to increased competition, while producers experiencing (greatly) inhibiting pass-through have been faced with markets tied up via marketing boards, marketing parastatals and *Caisse de Stabilization* schemes. There are of course exceptions: South Africa has liberalized its markets but appears to have inhibited the exchange rate pass-through. The perceived inhibiting effect occurred because for many crops, especially maize, prices were maintained above those on world market and liberalization resulted in a decline in prices to lower border parity levels. In several countries (Mozambique and Cameroon) the pass-through of one crop has been enhanced while for the other it has been inhibited. Mozambique appears to have enhanced the pass-through to cashew nut producer prices and inhibited the pass-through to cotton prices. The improved pass-though to cashew prices is the result of the removal of the raw cashew nut export ban and the reduction in the export tax. The inhibited pass-through to cotton prices is due to the continued rigidities of the administrative price controls set by the government through the Cotton Institute of

Mozambique. Likewise in Cameroon, the pass-through was enhanced for coffee producers while inhibited for cocoa and cotton producers. In 1994, Cameroon removed all quantitative restrictions on coffee exports and all coffee price controls were abolished. However, for cotton, SODECOTON continued to control cotton prices and marketing. The pass-through to cocoa prices in Cameroon is an interesting case. While cocoa marketing underwent significant reforms in 1994 the pass-though of currency devaluations to producer prices seems to have been inhibited during the 1990s. This could be explained by price interventions in the cocoa market, which maintained relatively high cocoa prices in 1990, even after the world price decline from the mid-1980s. Thus, changes in prices between 1990 and 1997 do not match the currency devaluation over the same period. If the period from 1993/94 is examine, then the pass-through has been more complete.

Cotton farmers seem to have experienced the lowest pass-through of currency devaluation to producer prices. Exceptions to this are Zimbabwe and Tanzania where cotton farmers experienced an improvement in the efficiency of the market with greater competition. The structure of these and other markets will be elaborated upon in the next section.

OBSERVATIONS FROM MARKETS AND PRICES OF EXPORT CROPS IN SUB-SAHARAN AFRICA

The state of the market and agricultural policies for the main export crops in Africa will receive more attention in this section. The analysis will examine the differences in producer incentives, focusing on the producer's share of the border price that farmers receive. Comparisons will be made among African countries as well as with other developing countries.

Cocoa

The recovery of world commodity prices, in the early 1990s, yielded a 14 percent increase in the real world price of cocoa between 1990 and 1995-97. The pass-through of this favorable price trend to cocoa farmers in Africa has shown mixed results and will be further examined in this section. The production of cocoa in SSA is highly concentrated with four West African countries producing 98 percent of the cocoa exports (Côte d'Ivoire [62 percent], Ghana [20 percent], Nigeria [10 percent] and Cameroon [6 percent]). An increase in the real domestic producer price was experienced by farmers in Nigeria, Côte d'Ivoire and Cameroon with average annual price increases of 6.8, 2.5 and 3.2 percent respectively, while in Ghana real producer prices were virtually stagnant (Table A6). In Nigeria and Cameroon, the producer's share of the f.o.b. price is greater than 75 percent which is comparable to the shares derived by cocoa farmers in other developing countries such as Malaysia, Indonesia and Brazil. In contrast, the shares enjoyed by cocoa farmers in Ghana and Côte d'Ivoire were marginally greater than 40 percent. These changing shares, as mentioned previously, are highly linked to the structure of cocoa marketing in these respective countries. Thus an analysis of these structures will provide some explanation on the sources of the differing market shares, as well as indication as to the constraints still faced by these farmers.

The marketing systems in Africa fall into three main categories: free market systems, marketing boards systems and price stabilization fund systems (*Caisses de Stabilisation*) (Schreiber and Varangis, 1999). The Marketing Board System is characterized by the existence of a parastatal with a monopoly on internal and external marketing. Pan-territorial and pan-seasonal prices are set by the boards or a higher governmental authority. The *Caisse de Stabilisation* is similar, with prices being administratively determined. The physical handling of the crop is conducted by private agents licensed by the *Caisse* and whose remuneration for these services is also determined by the *Caisse*. The purchasing and selling prices at each stage of internal commercialization and exports is fixed for the crop year. These systems attempt to stabilize prices with administrative price setting to insulate farmers from excessive fluctuations in international prices. Cocoa prices are set in relation to a long-run trend. In theory, export revenues derived when international prices are above the trend are supposed to finance a stabilization fund to cover losses when world prices fall below the trend. In practice this has not happened due to poor management of the stabilization funds by governments, increased difficulty in establishing the appropriate long-term price trends and political pressure on setting producer prices. Examples of this political pressure are evident in Côte d'Ivoire and Cameroon, following the financial difficulties experienced by the stabilization funds after the mid-1980s. World prices fell significantly lower than the mid-1970s and early 1980s levels, and governments were unable or unwilling to adjust purchasing prices to these lower levels, thus exacerbating the financial burden on these stabilization funds.

Free market system: Cameroon, Nigeria, Indonesia, Malaysia, Brazil.
Caisse de Stabilisation System: Côte d'Ivoire, (Cameroon until 1993/94).
Marketing Board System: Ghana, (Nigeria until 1986).

Table 4.3: Marketing Systems for Cocoa in Sub-Saharan Africa, 1997.

Functions	Free Market System	Caisse de Stabilisation	Marketing Board System
Legal Ownership of Crop	Trader, Exporters	Traders, Exporters	Marketing Board
Physical Handling of Crop	Traders, Exporters	Licensed Private Agents	Marketing Board
Domestic Price Determination	Market Forces	Caisse de Stabilisation (Government)	Marketing Boards (Government)
Taxation	Absent or Explicit	Explicit	Implicit
Marketing Costs & Margin	Low	Medium to high	High
Marketing Costs and Taxation	% of export price, 1995 *Indonesia -* 22 *Malaysia -* 9 *Brazil -* 28 *Cameroon -* 25 *Nigeria -* 13	% of export price, 1995 *Côte d'Ivoire-* 53	% of export price, 1995 *Ghana-* 49
Producer Prices	High	Medium to low	Low

Source: Adapted from Schreiber and Varangis (1999).

Countries under the free market system have enjoyed a higher share of the f.o.b. price (Figure 4.7). Prices are determined by market forces and there is no direct government involvement in the internal and external marketing of the crop. The pass-through of international prices to domestic prices is high but any volatility in international cocoa prices is, of course, also transmitted to farmers. In these systems, governments may retain the right to intervene if there is a perceived need to co-ordinate or regulate the actions in the system. However, this intervention is usually limited to quality control, taxation and a general system of monitoring and supervision.

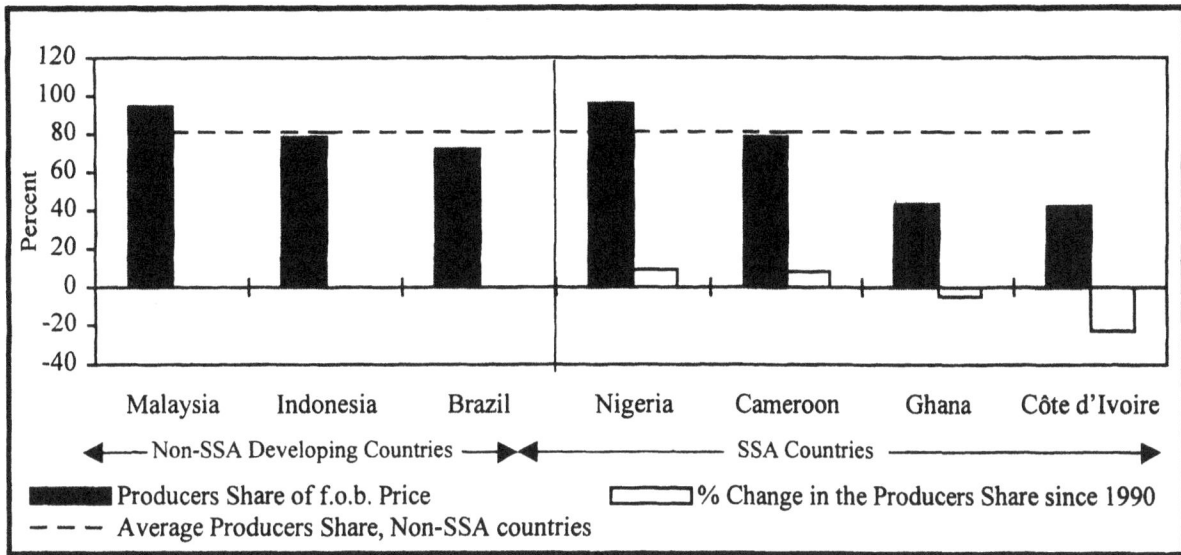

Figure 4.7: Cocoa Producer's Share of f.o.b. Price, 1995.
Source: Schreiber and Varangis, 1999, World Bank and IMF data.

The cost of marketing and taxation of the controlled systems is almost twice that of the free market system (Table 4.3). These costs are also strikingly evident in the substantially lower share of the f.o.b. price that cocoa growers receive in countries operating under the *Caisse* and Marketing board systems (**Côte d'Ivoire** and **Ghana**, Figure 4.7). In **Côte d'Ivoire**, prices are administratively determined by *CAISTAB (Caisse de Stabilisation et de Soutien des Prix des Produits Agricoles)* with the objective of stabilizing inter- and intra-annual prices and returns and reducing the price risks of market participants. A rigid system of controls and regulations is also managed by *CAISTAB*. It covers the entire marketing chain from the purchase of cocoa from farmers to actual export. This monopoly power essentially allows *CAISTAB* to determine the revenues of all participants in the cocoa marketing chain. A similar situation occurs in **Ghana,** with the Cocoa Marketing Board (COCOBOD) controlling all stages of cocoa marketing, from purchases at the farm-gate through exports and sale to domestic processors. COCOBOD's cocoa marketing activities are handled by separate subsidiaries: the Produce Buying Company (PBC) and the Cocoa Marketing Company (CMC). PBC is responsible for domestic procurement, storage and transport, while CMC manages all exports and sales to domestic processors[18]. Producer prices are fixed for the entire crop year. The set prices account for expected export prices, operating costs of COCOBOD and its various subsidiaries, the explicit tax and the farmers production costs.

Nigeria and Cameroon also have a history of state intervention in the cocoa sector. Prior to 1986, **Nigeria** experienced extensive government intervention. The Nigeria Cocoa Board (NCB), a national parastatal, set the producer price and held a monopoly on domestic procurement sales and exports. The deteriorating macro-economic environment, marked by rising domestic inflation, an increasingly overvalued exchange rate, and steadily worsening terms-of-trade for agricultural producers made cocoa production increasingly less profitable for farmers. In real terms, official producer prices halved between 1978/79 and 1984/85, and by 1986/87 farmers received less than 20 percent of the world price. In 1986 the government undertook widespread reforms. These included the adoption of a

---

[18] This system has been repeatedly changed between competitive private marketing and state-controlled monopsony.

more appropriate exchange rate and the abolition of the NCB. Price controls were abolished so were the licensing required by NCB. A subsequent surge in the number of buyers (400 after the abolition of NCB) created increased competition resulting in a significant increase in the farm-gate price for cocoa.

In **Cameroon** the parastatal *Office National de Commercialisation des Produit de Base* (ONCPB) controlled the marketing of cocoa until 1990. This system was similar to the*Caisse's*. Due to financial difficulties resulting from declining world prices and an increasingly overvalued exchange rate, substantial reforms were undertaken in the early 1990s with the elimination of ONCPB. Internal and external marketing were still partly controlled by a smaller parastatal company, *Office National du Cafe et du Cacao* (ONCC), continuing a stabilization fund. By 1994, again, financial pressure led to the adoption of an almost completely free marketing and export system. Producer prices are now market-determined and internal and external trade are open to everyone. At present there are about 200 registered buyers/exporters of cocoa (Schreiber and Varangis, 1999). The liberalization of the marketing system had a tremendous impact on farm-gate prices which increased from about 40 percent of world prices in 1994 to over 70 percent in 1995.

This description suggests a high correlation between the share of the f.o.b. price that cocoa producers receive and the extent of governments intervention in the industry. International experience suggests similar outcomes. The free market system is pursued by other developing countries such as **Indonesia** where the competitive cocoa marketing and distribution system has greatly improved the share of the f.o.b. price that cocoa farmers receive (Akiyama and Nishio, 1996). This, together with the availability of suitable land, low production costs, relatively good transport infrastructure, favorable macroeconomic policies and the smallholders' entrepreneurship have contributed to a phenomenal growth of Indonesia's cocoa output (26 percent a year between 1980 and 1994).

Coffee

Coffee production is less concentrated than the cocoa industry in SSA. Seven countries produce over 80 percent of SSA's coffee exports with no single country accounting for more than 20 percent of the export share. Of all the traditional export crops from Africa, coffee experienced the largest increase in the real world price between 1990 and 1995 with an annual (increase) growth rate of almost 10 percent. As a result of this favorable price trend, all main coffee producing nations in Africa experienced, on average, an increase in the real producer price of coffee (Table A6). Most of these countries also experienced an increase in the producers share of the f.o.b. price (Figure 4.8). This suggests that the pass-through of currency devaluations undertaken by these countries have been fully transmitted to coffee producers, particularly in Tanzania, Cameroon and Uganda. On average, the producer's share of the border price is higher for coffee producers than for cocoa and cotton producers. In some countries, these shares are comparable to those experienced by coffee producer's in other developing countries. Again, a brief description of marketing systems in these countries can provide an indication as to the constraints inhibiting the pass-through of border prices to producers. Both free market systems and marketing boards are present in the coffee industry in Africa.

Free market system: Uganda, Tanzania, Brazil, Indonesia, Kenya, Madagascar.
Marketing Boards: Côte d'Ivoire (until 1998/99).

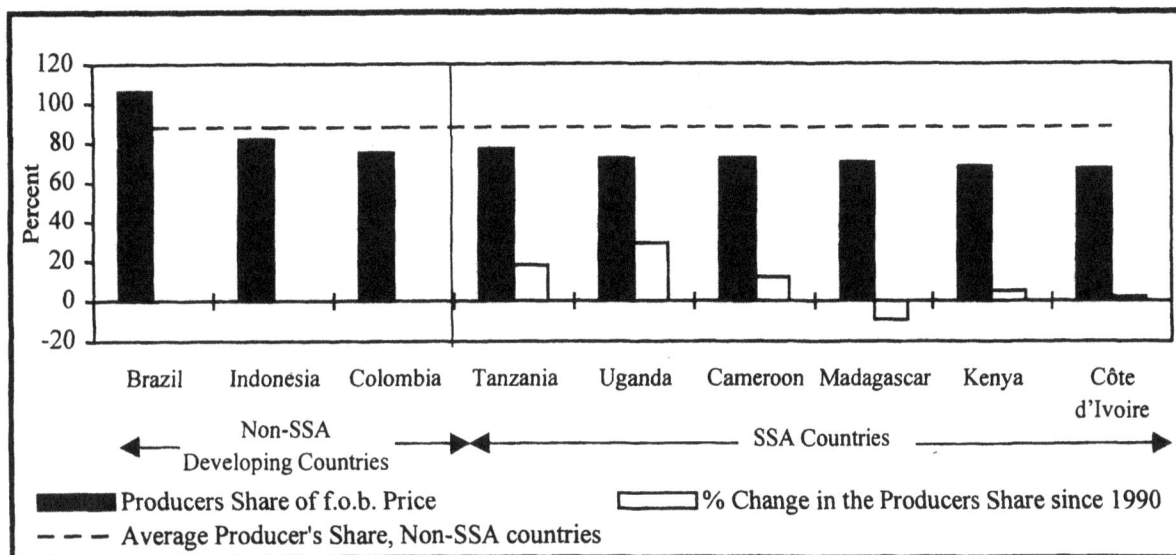

Figure 4.8: Coffee Producer's Share of f.o.b. Price, 1995.
Source: World Bank and IMF data.

Until 1993, the coffee sector in **Tanzania** was largely run by the Tanzania Coffee Marketing Board (TCMB), which controlled prices, marketing and exports. Restructuring of the TCMB began in 1993 with its functions confined to policies and regulation of the sub-sector. The internal purchasing of coffee was liberalized; this allowed the private sector to participate, thus improving competition in internal marketing. The ownership and management of coffee curing companies have also been liberalized by permitting private sector ownership. This opening up of the market has increased competition and improved efficiency, which has resulted in higher real producer domestic prices and enhanced pass-through of the currency devaluation to these prices.

In **Uganda** the coffee market has been liberalized with the encouragement of private sector participation in exports. The Coffee Marketing Board was replaced by a parastatal, CMBL, which does not have monopoly powers. The export tax on coffee was removed after 1991/92 but was reintroduced in 1994/5 and 1995/96 in the form of a stabilization tax. The intention was to curtail possible inflation owing to the rapid increase in the international price of coffee. In more liberalized markets, the number of private traders increased from zero in 1990/91 to 82 in 1994/95. These traders have taken over some of the export responsibilities from the coffee marketing board. The result has been a substantial increase in the real producer price and the producer's share of the f.o.b. price.

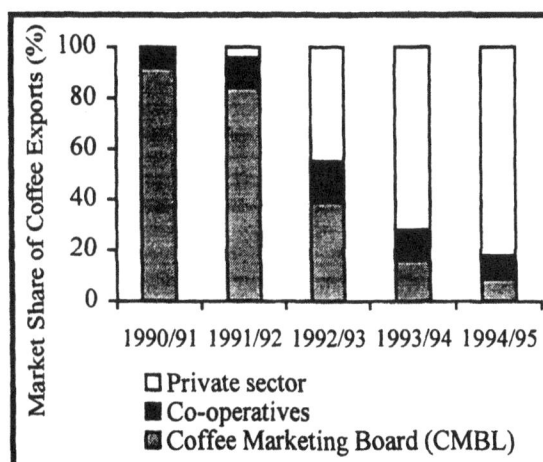

Figure 4.9: Changes in the Market Shares of Coffee Exporters.
Source: UCDA.

In **Cameroon,** the state marketing board, ONCPB, controlled coffee processors, marketing and export licenses. Following the currency devaluation in 1994, all quantitative restrictions on imports were removed and all price controls were abolished. A number of structural and regulatory reforms were implemented, which included eliminating the parapublic monopoly and control over coffee. There was some policy backsliding to reassert government controls with the reintroduction of coffee taxes (25 percent).

In the mid-1980s, **Madagascar** implemented a broad range of reforms in the coffee sector; taxes were removed and the state export monopoly was liberalized. During the early 1980s coffee producers received about 40 percent of the border price. By the early 1990s, farmers were typically receiving over 70 percent of the border price. In **Kenya**, the Coffee Board was responsible for selling coffee to private dealers and exporters. In October 1992, the sector was reformed and the Coffee Board ceased its involvement in coffee marketing, confining itself to regulating the industry. Regulation included licensing coffee growers, processors, and other marketing agents. The Kenya Planters Co-operative (KPCU) took over the export auction from the Coffee Board; however, farmers were also free to sell coffee by private treaty (outside the auction) as long as the price was higher than the auction price. The coffee auction is conducted in foreign currency with 50 percent of the foreign currency earned accruing to the seller of coffee.

In **Côte d'Ivoire**, coffee farmers have experienced significant fluctuations in producer prices. The share of the f.o.b. price received by farmers reached the lowest point in 1984/85 before rising sharply throughout the rest of the 1980s to 80 percent in 1992/93. It declined dramatically in 1994 as producer prices were not adjusted to reflect the exchange rate realignment and the increase in the international market price. The share then rose to around 60 percent in 1995/96 (Figure 4.8). These changing incentives have resulted in fluctuating production levels. The sluggish pass-through of world prices to producer prices is largely the result of both internal and external marketing. CAISTAB, a statutory monopoly (also responsible for cocoa marketing described in the previous section) played a dominant role in the marketing of coffee in Côte d'Ivoire. Its involvement inhibited private sector development, thus inhibiting efficiency in the marketing chain.

Cotton

Cotton exports from SSA have expanded significantly in the past decade resulting in an increasing share in total agricultural export earnings. Production of cotton is even less concentrated than that of coffee with ten countries (Mali, Côte d'Ivoire, Benin, Tanzania, Burkina Faso, Chad, Cameroon, Togo, Zimbabwe and Senegal) producing over 80 percent of the exports. Africa's share of the world cotton market has increased over the 1990s and currently accounts for about 15 percent of world exports. Most of this cotton is grown in West African countries.

The real world price of cotton lint (US$) has been more favorable over the 1990s than in the 1980s, increasing at 0.5 percent per annum between 1990 and 1997. All of the ten counties analyzed, except Chad, experienced an increase in the real domestic producer price of cotton with the largest increases enjoyed by cotton farmers in Zimbabwe and Tanzania. A closer examination of the share of the f.o.b. price that cotton farmers receive shows that all of these countries, except Zimbabwe and Tanzania, have experienced a declining share. This suggests that the devaluations of the domestic currencies have not been fully passed on to cotton farmers as higher prices. A closer examination of the marketing structures provides some explanation of these contrasts.

Free Market System:	Zimbabwe, Tanzania, India, Pakistan.
Parastatal System:	Togo, Côte d'Ivoire, Cameroon, Chad, Benin, Mali, Burkina Faso.
Marketing Board:	Zimbabwe until 1993.

The organization of the cotton industries in each of the francophone countries is very similar, with a dominant parastatal controlling cotton production. A recent study reviews the cotton policies in francophone Africa highlighting the functions of the parastatals, which are jointly managed by foreign shareholders (Pursell and Diop, 1998).

Table 4.4: Cotton Parastatals in Francophone Africa.

Functions	CIDT (Côte d'Ivoire)	SOFITEX (Burkina Faso)	CMDT (Mali)	SONAPRA (Benin)	SOTOCO (Togo)	Coton Tchad (Chad)	SODECOTON (Cameroon)
Input supply	Parastatal is the sole supplier; quantities and types of inputs are set to optimize processing capacity of cotton companies gins.						
Marketing	Parastatal is the buyer of seed cotton from farmers.						
Ginning	All cotton gins are owned and operated by parastatal.						
Transportation & distribution	Parastatal provides all transport for inputs from ports to farms, seed cotton from farms to gins, cotton seed from gins to mills or other markets.						
Producer prices	Pan-seasonal and pan-territorial seed cotton price announced before planting.						
Extension	Provide extension services, together with free new seed varieties.						
Credit	Supply input credits which are attached to the cotton price.						

Source: Adapted from Pursell and Diop (1998).

This system in francophone Africa (Table 4.4) has achieved a number of successes including significant growth in cotton production, high quality standards, good credit recovery, effective research and extension with a resulting increase in yields. This cotton success story in West Africa can be illustrated by way of a country example. **Benin** has experienced the greatest increase in cotton production (20 percent per annum since 1980 and 2 percent of this annual increase was from increased yields while 18 percent was from an expansion of area planted). This significant growth has been attributed to the impact of policies (Brüntrup, 1997). These have included both price and non-price factors combined in an integrated structure which links research, input supply, credit, organization of cotton commercialization and export, farmers cooperation and extension. Aside from the cotton price policy and the cotton package itself, Brüntrup (1997) also found that stagnating or declining food crop prices since the mid-1980s had an important influence on the cotton boom. Widespread adoption of animal traction, encouraged with access to credit, extension and commercialization services, has had a significant impact on labor productivity and cotton production.

Indeed, after an extensive review of the literature, Kelly *et al* (1999) suggest that input and output markets have served farmers best when there has been some degree of vertical co-ordination among input distribution, output marketing and credit functions, which lowers costs and improves loan repayment rates. The most successful examples of vertical co-ordination have been in sub-sectors producing industrial or export crops (as in the case of cotton in West Africa). In such cases, increased access to improved inputs and more reliable output markets stimulate productivity in food as well as in cash crops.

The key feature for the sustainability of these cropping systems is that they need to provide the appropriate incentives that make it profitable for farmers to sell their products through the scheme (assured largely by the *private sector*, with some government involvement to facilitate efficient and transparent markets). This in turn makes it profitable for the scheme to extend credit, inputs and other services that support small-holder productivity growth to the mutual benefit of the farmer and the scheme (Kelly *et al*, 1999). Where this has not been the case, the system often breaks down. Currently, price incentives provided to cotton farmers in west Africa are very low (Figure 4.10) in the range of 30 to 40 percent (in 1994/95 these prices were about half of those received by cotton farmers in Zimbabwe, Pakistan and India [these comparisons are for cotton of about the same quality]). These prices are offered by the cotton parastatals who manage stabilization funds with the idea of providing farmers with stable prices (in a similar manner to the *Caisse* system for cocoa). However, world cotton price fluctuations placed financial pressure on some of these funds and in some years many have needed financial assistance from aid donors.

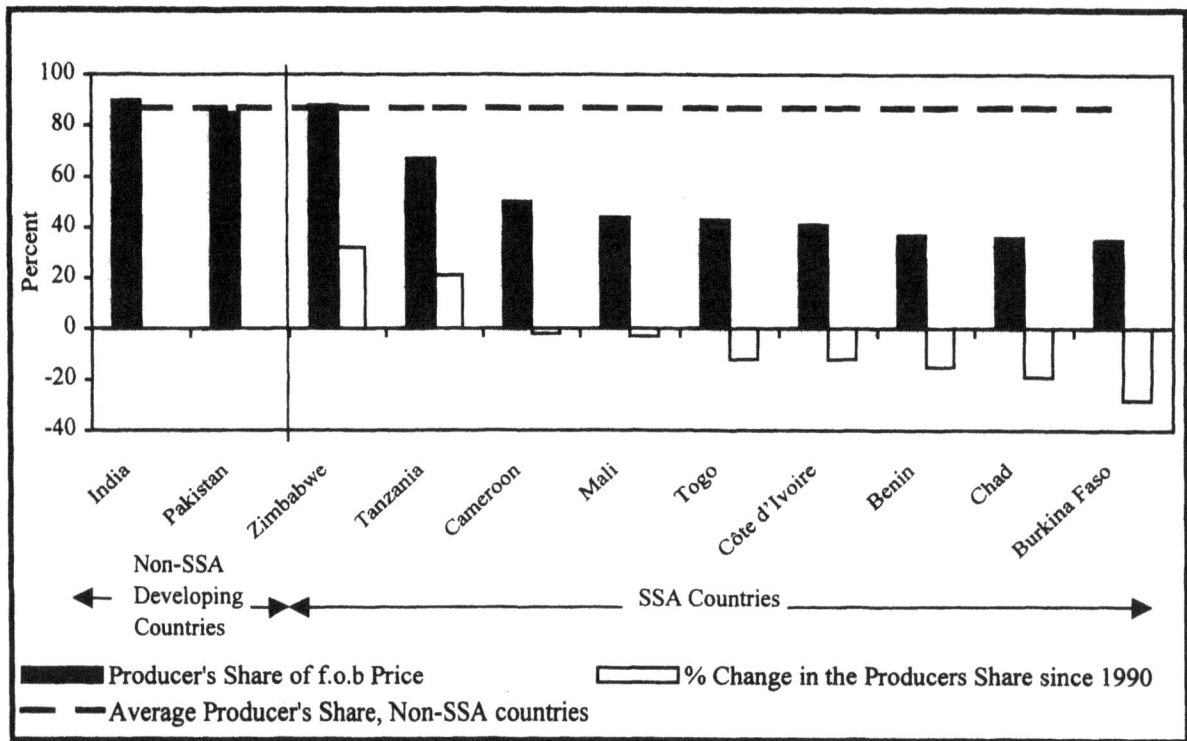

Figure 4.10: Cotton Producer's Share of f.o.b. Price, 1997.
Source: Purcell and Diop (1998).

The subsequent reorganization of many of these companies include a greater participation of private firms in transportation, input supply and ginning (ie: Benin and Togo). In the past, farmers were paid a pre-announced price minus the price charged to them for the inputs that had been supplied. The recent pricing arrangements for seed cotton include two payments to farmers: an initial price paid, supplemented by a second payment representing a share of the profits made by the cotton company during the season, or directly linked, by a formula, to the world price of lint (Pursell and Diop, 1998).

The extent of reforms varies greatly among countries, but the monopolistic structures have been maintained. Reform efforts are being discussed to raise competition at all levels and to harmonize cotton marketing and taxation policies within each union (UEMOA and CEMAC) to create a unified market who's size would be attractive to international agro-industrial investors.

In **Zimbabwe,** the cotton market has recently been liberalized. In addition, new traders are permitted, and the buying monopoly, the Cotton marketing board, has been downsized to become the CCZ (Cotton Company of Zimbabwe) and will eventually be privatized. Subsequently producer prices have gone up (Figure 4.10) significantly with increased competition in buying cotton which started in the 1993/94 season. In **Pakistan**, farmers receive 85 of the export price of cotton (lint). This high share is largely attributable to the effective role of traders in providing inputs and credit to small scale cotton farmers, and to the low costs of the competitive ginning, marketing and transportation activities. The large increases in production resulted from both increases in yields as well as from bringing new land under cultivation. In **India**, a large number of competitive seed cotton auction markets throughout the cotton growing regions ensure a market determined price.

Other Traditional Exports Crops

Other important traditional export crops are tobacco, sugar, groundnuts and tea, each accounting for over 5 percent of agricultural crop exports from SSA. A brief description will be provided on the price trends of the these crops across producing countries.

Tobacco accounts for about 9 percent of crop export earnings from SSA with a high level of concentration. Only two countries, Zimbabwe and Malawi, account for over 90 percent of these exports. Despite the decline in the real world price for tobacco, most producers in SSA have experienced real domestic price increases. This trend is a result of currency devaluations and an increase in the share of the world price received by farmers. The **Zimbabwe** tobacco industry has been largely free from interventions and prices reflect world price trends. 'Orderly marketing' is enforced through the auction floors (houses) with rules for bailing and grading. In general, production of flue-cured tobacco (commercial farmers) is expanding while the pattern for the other types of tobacco (varied types of producers) is more varied. International marketing of tobacco leaf is conducted by multinationals and there have been some concerns that these companies collude to form a buyers cartel thus creating distortions in transmitting the world price to farmers. The tobacco industry in **Malawi**, notably burley and flue-cured tobacco, was characterized by various production and marketing quotas at national and farm levels. National production quotas are set by a small number of international buyers and processors and, although all marketing takes place through auctions, they are characterized by a low degree of price competition. Production quotas were historically only allocated to estates until 1991/92. In 1994/95 smallholders were allocated about 10% of the quota which almost doubled in 1995/96. These production quotas are being phased out and replaced by production registration.

Sugar accounts for about 9 percent of SSA's agricultural exports. Three producers, Mauritius, Swaziland and South Africa, account for over 80 percent of sugar production. **Mauritius** supplies almost 40 percent of these exports. Farmers in **South Africa** and Mauritius enjoy about 90 percent of the f.o.b. price and the industry in these countries is largely free from government intervention. In Mauritius, sugar output is marketed through a central organization, the Mauritius Sugar Syndicate, responsible for the export and domestic sales of the country's production. Net proceeds are distributed to the large estates as well as to small farmers on the basis of the average realization price per ton. Mauritius sugar exports are shielded from the price fluctuations in the world sugar market by preferential access agreements with the European Union and the US, which secures a guaranteed price for sugar exports. These preferential trade agreements provide Mauritius with sugar prices much higher than the world price. In **Swaziland** the marketing of sugar is controlled by the Swaziland Sugar Association. Millers buy the sugar cane from farmers who in turn sell their output to the Association which then exports most of this sugar to the European Union and the US. Although farmers in Swaziland received a lower share of the f.o.b. price, this share increased in the 1990s. In Mauritius, Swaziland and South Africa the producer's share of the world price has increased.

Four countries (South Africa, The Gambia, Senegal and Zimbabwe) account for over 80 percent of SSA's groundnut exports with South Africa accounting for almost 50 percent of these exports. Prior to 1995, groundnuts in South Africa were marketed through a one-channel pooling system by the Oilseeds Board governed under the marketing act of 1968. This entailed strict government regulation over the production and marketing of groundnuts with the oilseeds board acting as a monopolist. The scheme was applicable to all groundnut products produced and marketed in South Africa. In 1995/96, the Oilseeds Board established a surplus removal scheme with a subsequent movement to a free market system. The board lost control over the production and marketing of oilseeds, and the one-channel marketing scheme was dissolved, resulting in increased competition among groundnut producers.

The groundnut market in **Senegal** was totally government controlled for several decades. In 1985 a set of reforms were implemented by the government that allowed greater private sector participation in the industry. The groundnut processing sector was, however, excluded from the reforms. In 1994, further reforms were implemented to remove import controls on vegetable oil and Government controls on consumer prices. Competition has subsequently increased due to growth in the informal oil processing sector and the production of confectionary groundnuts. However, these account for a small component of the total market for groundnuts and groundnut oil. Despite reforms, import protection was structured in a way to make competition with SONACOS (National Groundnut Oil Company) very difficult. As of 1998, tariffs of 26 percent and 48 percent were imposed in raw oil and refined oil respectively, thus protecting the domestic industry (SONACOS) for refined oil. As a result, competition in this sector continues to be inhibited.

Non-Traditional Export Crops

The previous analysis has centered on the 'traditional' export crops from SSA which has been the focus of many studies on African agriculture. A more recent study (Jaffee and Morton, 1995) attempts to complement this broader literature by focusing on high-value food products and raw materials. These include fresh and processed fish products, live animals, fresh and processed meat products, milk and other dairy products, fresh fruit and vegetables, processed fruit and vegetable products, tree nuts, oilseed and vegetable oils and spices and flavoring. High-value foods are produced by several African countries, particularly in Southern Africa.

Cut Flowers:	Kenya, Zimbabwe, Mauritius, South Africa, Zambia, Malawi
Citrus Fruit:	Swaziland, South Africa, Zimbabwe
Deciduous Fruit (Pineapples):	Swaziland, Côte d'Ivoire, South Africa
Pulses:	Ethiopia, Malawi
Live Animals:	Burkina Faso, Namibia, Lesotho, Niger, Somalia
Animal Products:	Ethiopia, Botswana, Niger, Namibia, Lesotho, Burundi, Somalia
Cashews	Guinea Bissau, Mozambique, Uganda
Vanilla:	Madagascar

These products have certain market characteristics and other properties that provide prospects for favorable expansion of future trade. Several of these have been highlighted by Jaffee and Morton (1995) as: high income elasticities of demand, greater potential for the development of domestic markets, intra-regional trade and more favorable international markets. These commodities also offer a wide scope for new product development and value adding activities. An examination of the price trends of these products reveals a long-term downward trend similar to the traditional exports. However, the downward trend has been less severe than the decline experienced by the traditional exports. The world price of citrus fruit, cashew nuts and vanilla have increased at 0.9, 0.7 and 0.6 percent per annum since 1961 (Table 2.4).

As with the traditional export crops, there has been government intervention in the marketing and price setting of these non-traditional exports. This again has undoubtedly eroded the producer's share of the border price. Rates of export concentration (Table 4.5) have also been high.

Table 4.5: African Examples of Export Concentration Among Private Firms.

Industry/Country	Share of Trade (%)	Number of Leading Firms
Horticulture		
Kenya Fruit/Vegetable	67	6
Senegal Fruit/Vegetable	81	5
Ghana Pineapple	63	6
Fish/Animals		
Nigeria Shrimp	74	3
Côte d'Ivoire Fish	>75	3
Somalia Cattle	70	3
Spices/Nuts		
Tanzania Cashew	64	3
Madagascar Vanilla	75	3

Source: Jaffee and Morton, 1995.

Examples of intervention include beef in Botswana, vanilla in Madagascar and cashew nuts in Tanzania. Beef in **Botswana** is marketed by the Botswana Meat Commission (BMC), a parastatal created by an Act of Parliament to provide a sales outlet for livestock farmers. This parastatal enjoys a monopoly over beef, lamb and mutton exports. However, on the domestic market, private butcheries, municipal abattoirs and livestock agents can compete. Although Botswana exports most of its beef to the European market through the Lomé Convention at a preferential price (about 40 percent above the world price) the export monopoly has eroded the producer's share of the f.o.b. price. By law the BMC has to make a profit and its gross revenues are taxed, this must surely be passed on to farmers as lower prices.

The vanilla sector in **Madagascar** provides another interesting case. Madagascar embarked on a program of structural reforms liberalizing its domestic and external trade, lowering its import tariffs, and gradually lowering and eventually abolishing all export taxes in 1992. The vanilla sector, however, was not included in the reform program. A Vanilla Stabilization fund was set up to protect domestic producers from excessive price fluctuations. However, the government continues to set all domestic and export prices as well as to allocate and set export quotas. Export taxes are persistently high. In 1990, vanilla growers received only about 8 percent of the actual export price. The stabilization fund collapsed in 1990 and now no longer provides a minimum price and a market outlet for their vanilla production. The producer's share of the border price has increased from about 8 percent in 1990 to 60 percent in 1995. In recent years, however, vanilla farmers have to complete with substitutes which have enjoyed an increased share of the world vanilla market. The Cashew nut market in **Tanzania** was liberalized in 1991 and private firms were permitted to undertake trade. The liberalization process was impeded but there were poor communications on new rules regarding trade. However, this move to a more market based system has resulted in a large increase in the producer's share of the f.o.b. price.

ENHANCING THE PRODUCER'S SHARE OF THE BORDER PRICE

Clearly, interventions in the market have distorted the pass-through of world prices to domestic producers. Analytically, the transmission of world agricultural prices to producers has been extensively analyzed. Mundlak and Larson (1992) using data for both developed and developing countries show congruence between world and domestic prices for agricultural exports. They also suggest that even though there are large cross-country variations in prices caused by policy biases, these do not prevent domestic prices from moving with world prices.

A more recent study by Morisset (1998) comes to a slightly different conclusion. His study examines the spread between the world and domestic price for seven commodities (beef, coffee, crude oil (fuel and gasoline), rice, sugar and wheat) in several OECD countries (Canada, France, Germany, Italy, Japan and the United States). He shows that the spreads have significantly increased over time thus inhibiting final demand. In the countries examined, commodity price transmissions have been asymmetric and price declines were not (or imperfectly) transmitted to domestic consumer prices. Upward movements, however, were fully transmitted and the spread between world commodity prices and domestic prices have almost doubled, on average, for the seven commodities between 1975 and 1994. As trade and tax policies or transport, processing and marketing costs only explained a small part of these trends, the study suggests that a large part of this explanation may lie with the role of international trading companies.

A recent African country case study (Lloyd *et al*, 1997) shows that in Côte d'Ivoire, domestic prices are insulated from world price shocks, with a large proportion of the shock being absorbed by the marketing agencies whose role was to primarily stabilize prices. In the case of coffee and cocoa in Côte d'Ivoire, Lloyd *et al*, (1997) show that the cost of stable prices for domestic producers is high. McIntire and Varangis (1998) derive a similar result; their findings show that in the absence of stabilization, farmers in Côte d'Ivoire would have received on average 29 percent higher producer prices throughout 1993-1997. This phenomenon does not seem to occur exclusively in Côte d'Ivoire but in most countries embarking on export price stabilization programs. While prices have been more stable with interventions, the risk benefits of the more stable prices do not appear to have outweighed the costs of the lower prices offered under the stabilization programs (Box 4.1).

The low share of the world price that farmers receive has been the focus of much work on international comparisons of agricultural policies. While this share has been eroded by heavy taxation of African farmers (Schiff and Valdés, 1992), other non-price policies such as those on transportation, marketing, distribution, storage and legislation have also influenced these prices. Isolating these individual effects has proved difficult and even calculations of nominal protection coefficients have their practical shortcomings (Gardner, 1995). In this section, comparisons will be made between the producer's share of the f.o.b. price with an attempt to explain cross-country variation in African countries in a more formal manner.

The distribution of the producer's share of the f.o.b. price in SSA shows a mode in the 70-79 percent class (Figure 4.11). However, most of the farmers receive shares between 40 and 69 percent of the f.o.b. price. This contrast indicates large differences in the cost of moving the product from the farm gate to the port. The question remains as to whether these high costs are justified or whether they can be reduced by removing distortionary policies and improving market efficiency.

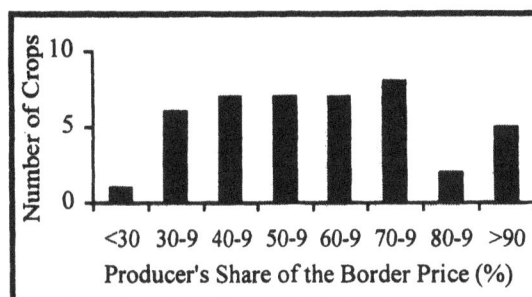

Figure 4.11: Distribution of Producer's Share of f.o.b. Price for Export Crops in Sub-Saharan African Countries.
Source: Table A7 in the Appendix.

An econometric analysis was used to explain the cross-country variation in the producer's share of the f.o.b. price. Cross-section data from 1996 was used in the analysis. The results in Table 4.6 suggest that differences in *infrastructure, isolation* from markets, *economic policies, volumes of crops traded* and the *type of crops grown* have a significant impact on the producer's share of f.o.b. price, suggesting complementarity between price and non-price factors.

Transport Infrastructure: Access to transportation infrastructure and the cost of transportation are significant constraints on African agriculture. Markets remain isolated, while poor road conditions and limited transport availability raise the costs of moving products to distant markets. The infrastructure variables used in the regression include the percentage of paved roads and road density, to differentiate between the extent of the road network and the quality of these roads (Table 4.6). The quality of roads appears to have a greater effect on the producer price/border price ratio than road density. These two variables did, however, appear to be collinear and thus road density was dropped from the later regressions (Model 2a and 2b). The positive sign on the variable suggests that a greater percentage of paved roads and a higher road density will increase the producer's share of the border price, reflecting lower transportation (transaction) costs.

Isolation: Similarly, the cost of moving crops from isolated areas to larger markets significantly reduces the producer's price share. Rural population density was used in the regression as a proxy for the extent of market isolation. The hypothesis being that in more isolated areas, market integration is weak and transaction costs for farmers are much higher than in areas or countries with higher rural population densities. The positive sign on these variables (Table 4.6) meets *a priori* expectations in that higher population densities (or lower levels of isolation) result in a higher producer's share of the f.o.b. price. Although positive, this variable was non-significant at the 5 percent level; this may simply reflect the inability, of the proxy variable used, to capture the isolation effects.

Box 4.1: Export Crop Price Stabilization.

The stabilization of commodity prices has been a common policy followed by many African governments. The critical question that is often posed when reviewing these programs is do the benefits from stable, but lower, prices exceed those from variable, but higher, prices ? One approach to examining this question has been provided by Newbery and Stiglitz (1981), which will be used in this evaluation. The Newbery-Stiglitz approach calculates the producer income effects of price stabilization policies by decomposing the effects it two components. The first is the risk premium, or stabilization benefit, indicating the monetary gain from more stable incomes where the risk benefit of perfectly stable prices is given by the equation.

$$RB = 1/2 * R\Delta\sigma_y^2$$

where R is the coefficient of relative risk aversion, and $\Delta\sigma^2_y$ is the fall in the squared coefficient of variation in income (Newbery and Stiglitz, 1981). The second component is the transfer benefit of perfectly stabilize prices and they represent this as:

$$TB = \frac{\Delta\overline{Y}}{\overline{Y}}$$

$\overline{Y}$ is the mean farm income before stabilization and $\Delta\overline{Y}$ is the change in income due to stabilization. Newbery and Stiglitz (1981) estimate what stabilization is worth to the farmer, that is, what sum of money, B, the farmer would be willing to pay for the stabilization scheme as

$$NTB = \frac{B}{\overline{Y}} = \frac{\Delta\overline{Y}}{\overline{Y}} - \frac{1}{2}R\Delta\sigma_y^2$$

The first term in the equation is the transfer benefit. The second term is the efficiency or risk benefit, the benefit of reducing costly risk to the farmer.

By applying the Newbery-Stiglitz formula to export crop data from 1975-1997 for individual African countries, the stabilization benefits were derived. Several assumptions were made for the analysis. The first was regarding the coefficient of relative risk aversion. Several studies have shown farmers to be only moderately risk averse (Binswanger, 1980,1981), however in African a more recent study (Brüntrup's (1997) analysis of cotton producer's in Benin) suggest farmers are highly risk averse. A range of values for R are used in the calculation of the risk premium, from 0.5 showing limited risk aversion to 1.5 showing higher risk aversion (R is the Arrow-Pratt measure of risk aversion). The second assumption is that in the absence of price stabilization, producer prices would experience volatility at similar levels to that of international commodity prices. Thus, the producer price volatility increases from zero (perfect price stabilization) to the level of the international prices.

For the calculation of the transfer benefits, it is assumed that in the absence of price stabilization, producer prices will rise to a share of the f.o.b. price similar to that in countries with open-market economies. The first three columns in the table below show the risk benefits (RB) of price stabilization. Taking the first row as an example, cotton producers in Mali, on average, would be willing to give up 2.5 percent of the average price for perfect price stabilization (assuming R=1). However without a stabilization scheme (government intervention) the average price that Mali cotton producers receive (46 percent of the border price) would likely increase to levels obtained in other developing countries that have no government intervention (ie: 85 percent of the border price). Transfer benefits are estimated to be –46 percent. In other words, in the absence of stabilization, farmers in Mali would have received 46% higher producer prices. Even if R, the coefficient of risk aversion is increase to 1.5 or even 2, suggesting that farmers are highly risk averse, the net transfer benefits of price stabilization are still negative.

Box 4.1: Export Crop Price Stabilization (*continued*).

Producer benefit of complete price stabilization, 1975-1997

Crop	Country	Perfect Price Stabilization				
		Risk Benefit			Transfer Benefit (TB)	Net Transfer Benefit (NTB using R=1)
		R=0.5	R=1.0	R=1.5		
Cotton	Mali	1.3	2.5	3.8	-46	-43
	Senegal	1.2	2.4	3.6	-45	-43
	Togo	1.5	3.0	4.4	-41	-38
	Côte d'Ivoire	1.0	2.1	3.1	-39	-37
	Burkina Faso	2.6	5.3	7.9	-37	-31
	Benin	1.7	3.3	5.0	-36	-33
	Tanzania	2.5	5.0	7.5	-33	-28
	Cameroon	1.3	2.7	4.0	-29	-26
	Chad	1.2	2.3	3.5	-26	-24
	Zimbabwe	0.7	1.4	2.1	-2	-1
Cocoa	Ghana	3.7	7.4	11.2	-57	-50
	Côte d'Ivoire	1.8	3.7	5.5	-32	-29
	Cameroon	1.7	3.4	5.2	-25	-21
	Nigeria	3.5	7.0	10.4	5	12
Coffee	Tanzania	4.4	8.7	13.1	-49	-40
	Madagascar	3.2	6.4	9.6	-45	-39
	Cameroon	2.0	4.1	6.1	-41	-37
	Côte d'Ivoire	3.0	5.9	8.9	-35	-29
	Uganda	3.8	7.6	11.4	-6	-2
	Kenya	3.8	2.6	11.4	9	11
Groundnuts	Zimbabwe	1.9	3.8	5.7	-58	-54
	Senegal	1.2	2.4	3.6	-53	-50
	The Gambia	1.6	3.2	4.9	-44	-41
	South Africa	0.7	1.3	2.0	-8	-6
Sugar	Swaziland	1.3	2.6	4.0	-35	-32
	South Africa	2.4	4.9	7.3	18	23
	Mauritius	0.8	1.6	2.5	34	36
Tea	Tanzania	2.8	5.6	8.4	-47	-41
	Malawi	2.1	4.2	6.3	-21	-17
	Kenya	1.3	2.6	3.8	-6	-3

The table shows several interesting results which deserve comment. Over the period examined the benefits from stable, but lower, prices which were offered by various stabilization schemes do not appear to have exceeded the benefits from variable, but higher, prices. Thus, the net transfer benefit has been negative. This result is particularly evident in counties and for crops were stabilization schemes have been put in place. The positive transfer benefit shown for some countries simply indicates that producer are receiving a higher share of the producer price than the average that was used in the calculations (85 percent of the border price), in many of these countries there has been an extended period of time with no-market interventions. In the case of South Africa and Mauritius, the received preferential trade agreements for sugar ensure exports price much higher than world prices. The model used in this brief overview is very simplistic and general equilibrium effects are not taken into account. Further work is need to refine some of these ideas.

In Africa, while there are certainly benefits from stable prices as the data suggest, these seem to have been outweighed by the costs imposed on farmers (lower prices) through the schemes implemented to achieve this stability.

Table 4.6: Explaining the Cross-Country Variation in the Producer's Share of the Border Price for Export Crops in Sub-Saharan Africa.

Dependent variable: Producer share of the border price for export crops

Market Efficiency Issues	Explanatory Variables	Model 1a		Model 1b		Model 2a		Model 2b	
		Coeff.	t-stat	Coeff.	t-stat.	Coeff.	t-stat	Coeff.	t-stat.
	Constant	49.806	8.645	66.780	6.403	48.236	8.557	65.608	6.338
Infrastructure & Isolation	Percentage of Paved Roads	0.187	1.084	0.271	1.747	0.308	2.685	0.382	3.583
	Road density	10.021	0.949	9.134	0.977				
	Rural population density	0.001	0.003	0.002	0.187	0.004	0.293	0.005	0.462
Economic Polices	Agricultural Market Interventions	0.582	3.759	0.508	3.597	0.601	3.920	0.526	3.761
	Real Interest Rate	-0.316	-1.474	-0.330	-1.652	-0.209	-1.147	-0.228	-1.341
Volumes supplied	Volume of crop produced	0.004	2.822	0.004	3.348	0.004	2.745	0.004	3.265
Crop specifics	Cotton			-0.772	-0.150			-0.542	-0.105
	Coffee			-18.516	-3.093			-18.533	-3.099
	Cocoa			-4.005	-0.577			-4.364	-0.630
	R²	0.62		0.74		0.61		0.73	
	F-statistic	7.97		8.08		9.41		8.98	
	Number of observations	36		36		36		36	

Dependent variable: Producer share of the border price for export crops (in logarithms)

Market Efficiency Issues	Explanatory Variables (in logarithms)	Model 1a		Model 1b		Model 2a		Model 2b	
		Coeff.	t-stat	Coeff.	t-stat.	Coeff.	t-stat	Coeff.	t-stat.
	Constant	3.336	5.718	3.347	6.285	3.019	5.769	3.178	6.529
Infrastructure & Isolation	Percentage of Paved Roads	0.035	0.712	0.058	1.269	0.067	1.571	0.076	1.873
	Road density	0.066	1.195	0.042	0.808				
	Rural population density	0.052	0.660	0.040	0.530	0.061	0.770	0.039	0.523
Economic Polices	Agricultural Market Interventions	0.011	3.901	0.010	3.783	0.012	4.344	0.010	4.200
	Real Interest Rate	-0.005	-1.756	-0.006	-2.195	-0.005	-1.624	-0.006	-2.098
Volumes supplied	Volume of crop produced	0.085	2.892	0.094	3.524	0.093	3.239	0.098	3.843
Crop specifics	Cotton			0.073	0.782			0.095	1.064
	Coffee			-0.256	-2.479			-0.254	-2.479
	Cocoa			0.068	0.563			0.063	0.518
	R²	0.59		0.71		0.57		0.70	
	F-statistic	7.03		6.98		8.03		7.88	
	Number of observations	36		36		36		36	

Table 4.6: Explaining the Cross-Country Variation in the Producer's Share of the Border Price for Export Crops in Sub-Saharan Africa (*continued*).

Dependent variable: Producer share of the border price for export crops

Market Efficiency Issues	Explanatory Variables	Model 1a		Model 1b		Model 2a		Model 2b	
		Coeff.	t-stat	Coeff.	t-stat.	Coeff.	t-stat	Coeff.	t-stat.
	Constant	58.060	8.920	80.228	7.190	57.907	9.097	80.275	7.326
Infrastructure & Isolation	Percentage of Paved Roads	0.337	1.935	0.425	2.697	0.366	3.305	0.451	4.426
	Road density	2.272	0.212	2.074	0.218				
	Rural population density	0.002	0.176	0.008	0.675	0.002	0.239	0.008	0.767
Economic Polices	Agricultural Market Interventions	-21.585	-4.045	-20.049	-3.897	-21.922	-4.372	-20.398	-4.247
	Real Interest Rate	-0.091	-0.398	-0.113	-0.521	-0.063	-0.342	-0.085	-0.492
Volumes supplied	Volume of crop produced	0.002	1.580	0.003	2.315	0.002	1.593	0.003	2.355
Crop specifics	Cotton			-5.682	-1.096			-5.722	-1.125
	Coffee			-19.243	-3.324			-19.227	-3.381
	Cocoa			-7.919	-1.138			-8.102	-1.193
	R²	0.64		0.75		0.64		0.75	
	F-statistic	8.6		8.7		10.7		10.2	
	Number of observations	36		36		36		36	

Dependent variable: Producer share of the border price for export crops (in logarithms)

Market Efficiency Issues	Explanatory Variables (in logarithms)	Model 1a		Model 1b		Model 2a		Model 2b	
		Coeff.	t-stat	Coeff.	t-stat.	Coeff.	t-stat	Coeff.	t-stat.
	Constant	3.550	6.170	3.586	6.540	3.294	6.296	3.385	6.631
Infrastructure & Isolation	Percentage of Paved Roads	0.057	1.158	0.085	1.795	0.085	2.077	0.108	2.651
	Road density	0.058	1.058	0.053	1.004				
	Rural population density	0.061	0.816	0.083	1.130	0.070	0.924	0.086	1.168
Economic Polices	Agricultural Market Interventions	-0.364	-4.168	-0.328	-3.971	-0.391	-4.668	-0.349	-4.018
	Real Interest Rate	-0.003	-1.124	-0.005	-1.526	-0.003	-0.975	-0.004	-1.361
Volumes supplied	Volume of crop produced	0.048	1.734	0.055	2.079	0.052	1.911	0.058	2.237
Crop specifics	Cotton			-0.052	-0.529			-0.033	-0.340
	Coffee			-0.285	-2.754			-0.285	-2.758
	Cocoa			-0.027	-0.211			-0.040	-0.313
	R²	0.61		0.71		0.60		0.69	
	F-statistic	7.61		6.06		8.87		7.50	
	Number of observations	36		36		36		36	

Percentage of paved roads measures the percentage of roads in a country that are paved; road density is measured in km of road per sq. km area; rural population density is measured as the number of rural people per square kilometer; agricultural market intervention is measured as the recent change (1990-1995/97) in the producer's share of the f.o.b. price in a particular country - an increase signifies an improvement in the country specific crop policy; volume of crop produced is measured in millions of tons. The source of these data are World Bank Development Indicators, 1998, IMF, and FAO. The repeated regressions (in the second set of tables) used a different variable as the proxy for agricultural market intervention. A dummy variable was used having the value of 1 in markets with government intervention and 0 in markets with no-government intervention.

Economic policies: Both sectoral and macroeconomic policy variables were used in the regression. Two proxy variables were used for agricultural market interventions, the first was the recent change in the producer's share of the f.o.b. price, the second a dummy variable (1 for government market intervention and 0 for no government intervention). For the first proxy, an increase in this share was taken to represent an improvement in economic policy while a decline was taken to represent regress. While this may be a generalization, there appears to be a strong correlation between the two (see description in previous sections). These variables both had the most significant effect in the regressions (Table 4.6), suggesting that in markets where governments intervene producers receive a lower share of the border price. When the dummy variable is used, the results suggest that, on average, producers receive a share of the border price which is over 20 percent lower than farmers receive in markets where governments do not intervene. Macroeconomic policy can have a direct effect on producer prices through exchange rates and money supply and an indirect effect through real interest rates. Private sector entry into markets and their subsequent investments (storage and transportation) is influenced by, among other things, the real interest rate. Excessively high rates of borrowing may prevent investment and thus inhibit competition. The real interest rate was included in the regression as an explanatory variable having a negative coefficient, thus suggesting that an increase in real interest rates reduces the producer's share of the border price (possibly through a reduction in private sector investment).

Volumes supplied: In many African countries the volumes of export crops produced are small and economies of scale in trade are not realized. A consistency in volume supplied is required so that the private sector can enter the market and, if this is not achieved, competition will not develop and the prices that the farmers receive will remain low. A variable for the volume of crops produced was included in the regression and proved to be highly significant and positive suggesting that an increase in the volumes produced by a country will result in a larger share of the f.o.b. price for producers.

Crop specifics: Finally, dummy variables were included for three specific crops – cotton, coffee and cocoa (these crops account for over 50 percent of agricultural exports from Sub-Saharan Africa). Coffee seems to be the only crop where producers receive a significantly lower producer's share of the f.o.b. price, estimated to be about 19 percent lower than the average. This, however, may simply reflect a failure to fully account for the transformation of coffee from the farm gate to the port (raw beans vs. washed beans). The results suggest that only a small portion of the cross-country variations can be explained by the difference in crops grown.

70

ENHANCING THE COUNTRY'S SHARE OF THE WORLD PRICE

A comparison between the border price of a country and the world price reflects the extent to which the world price has been transmitted to the border price. A higher border price/world price ratio could be due to factors such as lower international transportation costs or a greater external bargaining power. The distribution of these shares is shown in Figure 4.12. Most of the crops analyzed had border prices in excess of 80 percent of the world price.

Again, an econometric approach is used to explain the cross-country variation in the border price/world price. Cross-country data from 1996 were used in the analysis. The regression results in Table 4.7 suggest that differences in *volume of exports traded*, *distances from markets* and *type of crop exported* all impact on the country's share of the world price. Agricultural market interventions did not appear to have influenced the countries share of world price.

Volume of exports: The export share of world trade was included in the regression to reflect a countries bargaining power on world markets. The larger the share, the greater bargaining power a country is likely to have. Indeed, the variable had a significant effect on explaining the variation in a country's share of the world price. The greater the share of world trade, the higher the price share a country received for its exports.

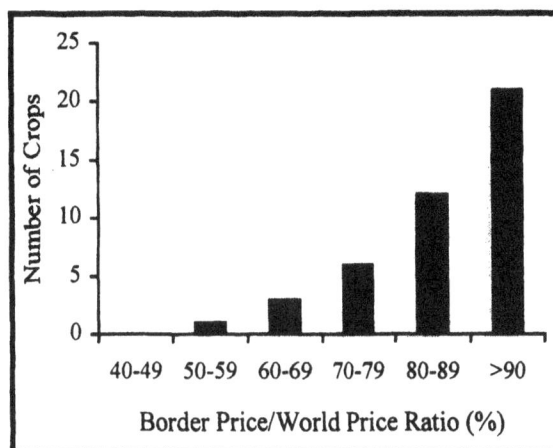

Figure 4.12: Distribution of the Percentage of the Border Price to the World Prices for Export Crops in Sub-Saharan Africa.
Source: World Bank and IMF data.

Table 4.7: Explaining the Cross-Country Variation in the Border to World Price Ratio for Export Crops in Sub-Saharan Africa.

Market Efficiency Issues	Explanatory Variables	Dependent variable: Border - World Price Ratio for Export Crops							
		Model 1a		Model 1b		Model 2a		Model 2b	
		Coeff.	t-stat	Coeff.	t-stat.	Coeff.	t-stat	Coeff.	t-stat.
	Constant	83.37	**15.82**	84.78	**4.13**	4.116	**26.814**	4.285	**11.980**
Volume supplied	Value exported	0.02	1.39	0.04	1.84	0.072	**2.155**	0.055	1.362
Quality	Export share of world trade	0.56	1.35	0.75	1.66	0.031	0.943	0.088	**2.007**
Distance from markets	Landlocked countries	-5.60	-0.70	-9.67	-1.16	-0.101	-1.242	-0.165	**-1.988**
Policy	Agricultural Market Interventions	-2.35	-0.34	-8.23	-1.02	0.033	0.475	-0.056	-0.677
Crop specifics	Cotton			-13.51	-1.44			-0.224	**-1.914**
	Coffee			2.55	0.26			0.015	0.144
	Cocoa			6.48	0.52			0.076	0.613
R^2		0.21		0.33		0.29		0.42	
F-statistic		2.88		2.35		3.12		2.85	
Number of observations		36		36		36		36	

In models 2a and 2b, the dependant and explanatory variables are in logarithms.

Distance from markets: A dummy variable (1 = landlocked, 0 = not landlocked) was included in the regression to represent cross-country variation in costs (transportation). The variable had a negative coefficient in the regression, suggesting that landlocked countries receive a lower share of the world price.

Type of crops exported: Dummy variables for the three dominant crops: cotton, coffee, and cocoa, were included in the regressions. The only significant dummy variable was for cotton suggesting that cotton receives a lower share of the world price than other crops exported from Sub-Saharan Africa. Again, caution should be used in interpretation as this could simply reflect different product comparisons at the border and on the world market. The same dummy variable for government intervention used in the previous regression was included. The variable was non-significant suggesting that in markets where governments intervene higher prices on world markets for domestic product have not been realized.

The argument that greater government intervention (low producer's share of the border price) allows higher border prices to be realized by a country due to the improved market (bargaining) power with centralized exporting agent, does not seem to have much weight at an aggregate level (Figure 4.13). As discussed in previous sections, a lower producer's share of the border price reflects high government intervention, but this intervention does not assure producers a higher price on world markets (the best fit line in Figure 4.13 is almost horizontal).

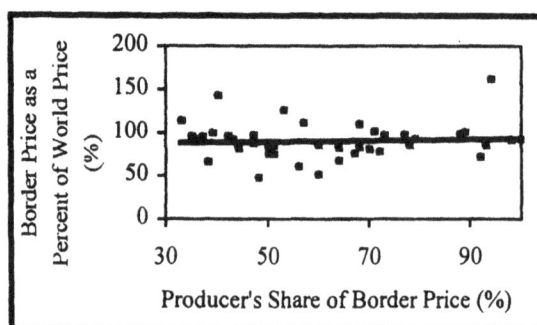

Figure 4.13: Relationship Between the Producer/ Border Price Ratio and the Border/World Price Ratio. *Source:* Table A7 (all crops and countries).

At a more disaggregate level, interventions in cocoa and coffee markets yields higher prices on world markets. However, this benefit is eroded by the lower share of the f.o.b. price that these farmers receive as a result of the intervention (Table A6).

AGRICULTURAL TAXATION AND PUBLIC RURAL INVESTMENT

As discussed, African agriculture has been subject to high taxation (both explicit and implicit) and many of the instruments used to extract resources from agriculture, such as the marketing boards, have generated little government revenue and have led to high deadweight losses. As this excess taxation is dismantled, the question arises as to how should the sector be taxed in the future ? It is obvious that agriculture, as the major sector in most African economies (accounting for 35 percent of the continents GDP), will have to continue to contribute to government revenues, ie: it will have to be taxed. In answering this question the key principles to be applied to future agricultural taxation are non-discrimination, minimization of negative efficiency impacts, effectiveness of fiscal capture, and capacity to implement (Binswanger, Townsend and Tishbaka, 1999).

Agricultural taxation, where possible should be integrated into general value added, profits, income and wealth taxation. Efficiency losses must be minimized via the minimization of output and input taxes. Consumption taxes (such as sales and value added taxes) have the advantage in that they don't effect the efficiency of production. Although consumption taxes have become a more popular form of taxation in Africa, administrative capacity to implement seems to remain limited in many Sub-

Saharan African countries. The capacity to implement these tax systems needs to be strengthened and, in many cases, will have to be built up over many years during which little revenue will be generated. In this context, export taxes may be justified until the administrative capacity is developed. It appears that where export taxes are justified as substitutes for income tax in Africa, rates should be reduced substantially. The same its true for input taxation.

As mentioned earlier, the high levels of taxation have had a crippling effect on agricultural growth. These effects would not have been as extreme if more of the revenues were reinvested into public services and rural infrastructure (which was shown in this chapter to have a significant effect on agricultural incentives). Indeed countries in East Asia provide examples of high extraction from rural areas combined with significant public investment and growth (Karshens, 1998).

In Africa public rural investment has been inhibited by the highly *centralized* political, fiscal and institutional systems for rural development that the post-colonial regimes have established. This focus on centralization occurred for many reasons, including the desire for political integration of fragile nations and the dominance of the state led development and planning ideologies of the time (Manor, 1999). In a recent research program of the World Bank on decentralization and rural development (McLean *et al*, 1999), scores were developed from a detailed analysis of the institutional arrangements for decision making and resource allocation in six important sectors of rural development (rural primary education, rural primary health care, rural roads maintenance, agricultural extension, rural water supply and forestry management which allow for the characterization of rural development institutions). (Since these scores are based on an analysis of the decision making powers at different administrative levels [e.g school, district, administrative region, state, central government], they are free of biases associated with country size).

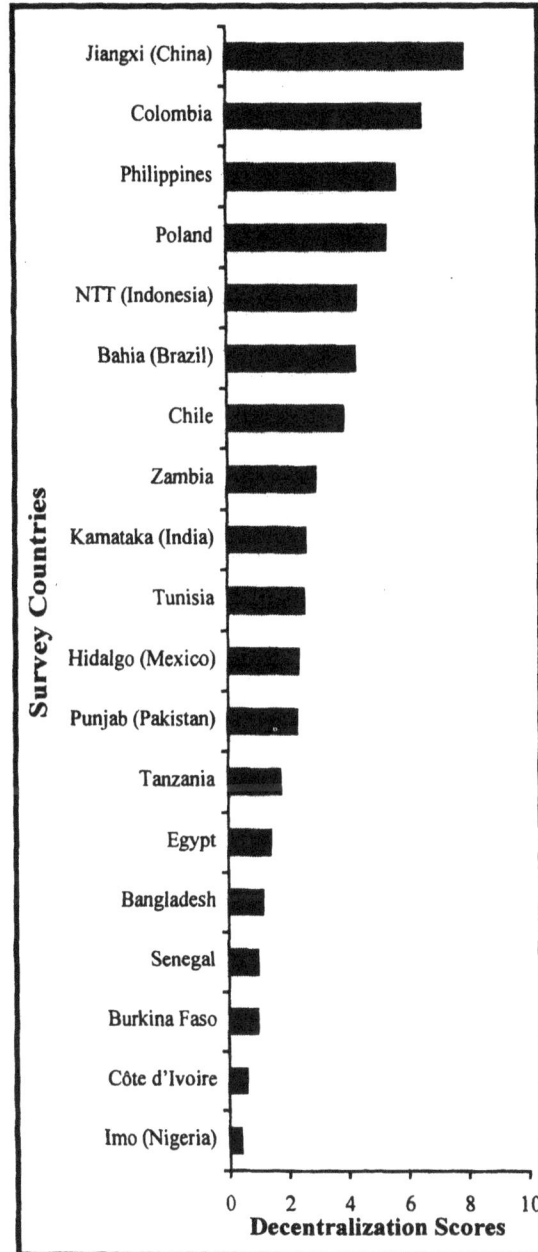

Figure 4.14: Rural Service Decentralization in Nineteen Countries in the 1990s.
Source: McLean *et al*, 1998.

It is clear from Figure 4.14 that the African countries included in the sample had the most extreme centralized institutions for rural development during the first half of the 1990s. These high levels of centralization inhibited the local level development of institutional capacity, limited local

73

resource mobilization, undermined accountability of development programs to local populations and inhibited their participation (McLean *et al*, 1998, Parker 1995). The inhibition of local initiatives was further aggravated by the lack of democracy in most of the countries, and the discouragement or even suppression of voluntary private associations.

The effectiveness and allocation of rural public investment will require a more decentralized approach. There has indeed been some progress in Africa of decentralized rural service delivery and public investment. Deconcentration of administrative and implementation responsibilities have become features of many sector investment programs for these services. Moreover, many countries have re-introduced political bodies at local levels (e.g. Tanzania, Guinea, Uganda, Ethiopia, Ghana, Zimbabwe). Progress in decentralization and devolution of resources and responsibilities however, remains limited to a few countries such as Ethiopia, Uganda and Ghana.

SUMMARY AND CONCLUSION

African countries have adopted and implemented widespread reforms in their export crop sectors. State control on pricing and exporting has been reduced particularly in eastern and southern African countries. In West Africa, the marketing systems of cotton and cocoa receive continued intervention. In some of these countries the markets are gradually opening up to increased competition.

- Between 1990 and 1995/97, there has been an increase in the real domestic prices for export crops in fifteen of the nineteen African countries analyzed. This improvement is the result of favorable world commodity prices, devaluation of domestic currencies and, in some countries, an improvement in sectoral policy.

- Since 1990 there have been large contrasts in the extent of exchange rate pass-through to producer prices. Many countries in Sub-Saharan Africa continue to inhibit this pass-through, particularly evident for cotton, cocoa and to a lesser extent coffee in West African countries. The exchange rate pass-through to farmers has been enhanced policies (real domestic price increase exceeding those suggested by the currency devaluation) in markets with open trade due to increased competition and market efficiency.

- Similarly the producer's share of the f.o.b. price remains low and in some African countries this share declined during the 1990s. The cross-country variation in the producer's share can largely be explained by differences in infrastructure (quality and quantity of roads), economic policies (both macroeconomic and agricultural) and the volumes of crops traded.

- One justification used for continued state price control is to stabilize domestic export crop prices. Observations from African export markets suggest that these more stable prices are achieve at great cost to the farmer. The risk benefits of more stable prices do not appear to have outweighed the costs of the lower price offered to farmers under the stabilization programs.

- Another argument in favor of continued or even greater government intervention is that countries are able to realize higher border prices when exports are centralized due to the improved bargaining power on world markets. Evidence suggests that export prices received in countries that opened up the export channels to greater competition are not significantly different from those with centralized marketing systems. However, one benefit from these centralized systems is that it provides provide a platform to monitor export quality.

Several government actions (related to price and non-price factors) can help to improve the income of export crop farmers in Africa. There seems to be significant scope to improve the exchange rate pass-through to domestic producer prices. Raising the producers' share of the border price could be achieved by improving the quality and quantity of the road network in individual countries (reducing transport and transaction costs), enhancing economic policies (removing high taxation), fostering production to ensure consistency and volume of crops traded. Many of these improvements will encourage private traders to enter into these markets, thus increasing competition and market efficiency.

Past taxation instruments, such as marketing boards, have generated little government revenue and have led to high deadweight losses. It is, however, obvious that agriculture, as a major sector in many African economies, will have to contribute to government revenues. Even though African agriculture was highly taxed there has been no extended period of active public investment in agriculture and rural development, a trend which needs to be reversed if Africa is to realize is agricultural growth potential. In the future, the key principles that need to be applied to agriculture taxation are nondiscrimination, minimization of negative efficiency impacts, effectiveness of fiscal capture and capacity to implement. This taxation needs to be accompanied by higher levels and more effective allocations of agricultural and public rural investment.

5.   FOOD CROP POLICIES, PRICES AND MARKETS

LIBERALIZATION OF GRAIN MARKETS IN AFRICA

Intervention in the marketing of food crops, like export crops, has been widespread in Africa. Under the Malthusian view held by many African governments, food self-sufficiency strategies were extensively adopted. Marketing boards for most crops were developed and food production was subsidized by policies consisting of cheap credits and input subsidies. Governments purchased grain crops at higher prices than those prevailing on world markets while selling them to consumers at subsidized prices. These subsidies, together with little effort at cost recovery, caused marketing board costs to escalate. Operational inefficiencies also exacerbated the financial burden of these interventions. These rising costs, unreliable input deliveries and crop payment, the existence of parallel markets resulting from pricing policies, and instability of purchases and sales increased pressure for reform (Box 5.1). These factors were fueled by an increasing pessimism about the motives and results of state intervention in agricultural markets and the optimism of the private sector to efficiently organize these markets (Lipton, 1991; Jones, 1994).

Following the initial failed attempts to improve the performance of these state marketing boards, the liberalization of parastatal food marketing systems became central to economic adjustments in Sub-Saharan Africa. A recent World Bank study (World Bank, 1994) shows that governments in fifteen of the twenty eight countries analyzed imposed major restrictions on purchases and sales of food crops. In an additional five countries there was limited intervention by government buying agencies. The remaining countries had no intervention except in food security stocks. The extent of the interventions has varied across crops. Markets for roots and tubers, predominantly grown in West African countries, have experienced less government interference than that imposed on maize and rice. The World Bank (1994) report shows that by 1991-92, after widespread adoption of adjustment programs, only two countries, Kenya and Zimbabwe, had major restrictions on maize purchases and sales. During the 1990s there has been backsliding on a number of these policy reforms.

The common reform package for maize marketing, the dominant staple crop grown in Africa, has included moving farm gate prices towards export or import parity; announcing administered prices in a more timely fashion in relation to the planting times of crops on farms; speeding up payments to farmers; eventually liberalizing prices altogether; relaxing maize movement controls and other restrictions on trade; and restructuring parastatal maize marketing companies (Donovan, 1996). In many countries the reform process has been slowed by the social and political sensitivity of food prices together with the recurrent droughts, particularly in Southern Africa. Concerns about food insecurity in the region has induced many governments to retain grain parastatals, managing strategic grain reserves and to continue to intervene in grain trade both on domestic and international markets. A summary of the stages of reforms in several African countries is summarized in Table 5.1.

Box 5.1: Assessment of Grain Market Reforms in Eastern and Southern Africa.

Factors Leading to the Reform of Grain Marketing Systems in Eastern and Southern Africa	Assessment of Market Liberalization
Some of the major factors driving marketing boards to move towards reform have been summarized by Jayne and Jones (1997). These factors include high marketing board costs, unreliable state systems for input delivery and crop payment, pan-territorial pricing, which encouraged parallel markets, instability in marketing board purchases and sales and limited market access to small-holders. • *Costs of the marketing boards escalated* These costs were manifest at two levels: first, the government enforcement on the board to carry out activities without allowing it to fully recover costs (ie: pan-territorial pricing increased the share of grain delivered to the board by small-holders in remote areas) and second, the costs of operational inefficiency. • *An increasingly unreliable state system of input delivery and crop payments.* Small-holder credit repayment also became problematic in many cases. • *Increased switching to parallel markets.* Farmers near urban demand centers who were implicitly taxed through pan-territorial pricing increasingly resorted to parallel markets and/or switched to other, uncontrolled crops. • *Increased instability of marketing board purchases and sales increased the fiscal demands made by the marketing system.* This resulted from the increasing proportion of maize sales by small-holders generally on poorer land with less reliable rainfall. • *Inhibiting small-holder market access.* Controls on private grain movement restricted small-holders direct access to urban markets. These also inhibited the flow of grain from surplus to deficit areas. • *Suppressed or imposed additional costs on parallel trading and processing channels.* These channels often served the interests of both producers and consumers more effectively than those of the official state apparatus.	• *Reduction in the cost of marketing food to grain-deficit rural areas.* Inflated costs of controlled marketing, rent seeking behavior and the controlled flow of grain raised consumer prices. The legalization of inter-district grain movement has reduced the difference between prices realized by producers and those paid by consumers. The reduction in marketing costs has been achieved primarily by expanding the role of small-scale trading and milling networks in fulfilling the residual grain needs of rural households. • *Changes in urban consumption patterns and improved access to food by low income consumers.* Urban maize milling in each country was dominated by several large registered firms using roller mill technology. After reforms were put in place, the large scale millers swiftly lost a major part of their markets to small informal hammer mills whose number rapidly expanded. Widely viewed during the control period as a product having negligible demand, whole maize meal now accounts for 40-55% of total urban meal consumption in Zimbabwe, Kenya and Zambia. Liberalization led to widespread reduction in food costs by removing some of the barriers on private trade. • *Limited supply response to food market liberalization.* Throughout the 1980s and up to the initial reforms, official producer prices exceeded export parity prices in the major production regions of Southern Africa. As these producer prices were reduced grain production has declined. In almost all countries, a large proportion of small-holders benefited from the transport subsidies inherent to the boards pan-territorial pricing structure. The removal of government controls on grain trading, however, has to some extent mitigated the adverse effects of declining state marketing subsidies associated with structural adjustment. Lower grain processing costs made possible through liberalization have reduced the wedge between producers and consumer prices. In some cases, price spreads between surplus and deficit regions have declined. • *Gradual movement of region to structural food deficits.* Even in a normal year, a large proportion of rural households are actually net buyers of grain. This indicates that the effects of the reforms on food security will depend on the ability of the emerging private sector to reduce costs of food to the grain deficit and generally poor areas. • *Impact of liberalization of marketing board deficits.* To date, there has been limited progress towards establishing more flexible pricing strategies that allow the state to respond to, or influence at the margin, prevailing prices in private trading channels. *Source: Jayne and Jones (1997)*

Table 5.1: Summary of Maize Marketing Policies in Sub-Saharan Africa.

Country	Pre-reforms	Post-reforms
Tanzania	Pre 1984 – Maize producer prices determined as residual after deduction of cooperative and National Agricultural Products Board costs. - Pan-territorial and pan-seasonal price setting. 1981 - Regional price setting introduced.	1984 - Maize flour price decontrolled, controls on grain movements relaxed. 1985 - Maize grain price decontrolled. 1986 - Maize flour subsidies removed. 1987 - Official producer prices to be regarded as minimum cooperatives prices, movement restrictions abolished. 1990 - All restrictions on purchase of grain by traders eliminated.
Zimbabwe	1970s (late) – Consumer and producer subsidies increased. Expansion of pan-territorial pricing, taxing commercial maize producers. 1985 - Consumer subsidies phased out, but reintroduced from 1991-93.	1987 - Producer price allowed to decline in real terms, then rise after 1992 drought. 1991/2 - Phased elimination of controls on trade between smallholder areas. 1993 - GMB monopoly-seller status restricted to large mills, then eliminated (1994). External trade still controlled by GMB. Maize meal subsidies abolished, consumer prices decontrolled.
Malawi	Pre 1987 – Pan-territorial pricing encourage production in Northern Districts. ADMARC subsidized maize producer prices and fertilizer.	1987 – Premiums set for maize delivered to main depots. Target of zero loss on ADMARC maize account. 1995 – Removal of price setting but maize price bands were established.
Kenya	Pre 1988 – Increase in the National Cereals and Produce Board (NPCB) purchase price. NCPB margins squeezed as imported maize sold at subsidized prices.	1988 – NCPB price margin narrows, limited unlicensed trade allowed. 1991 – Further relaxation of district trade limits. 1992 – NCPB unable to defend ceiling price, movement restrictions tightened. 1993 – Unable to defend floor price. 1994 – Liberalization of internal trade; continued restrictions on external trade.
South Africa	Pre 1987 – Single channel fixed price scheme, prices set on a cost plus basis.	1987 – Fixed producer price replaced with advanced price and supplementary payment to ensure maize account breaks even ('pool pricing scheme'). 1993 - Direct farmer-to-miller trade permitted, subject to levy payment and approval. 1995 – Maize price controls removed, restrictions on private maize purchases and sales removed, imports liberalized with zero tariff.
Mozambique	1981 - Parastatal set up – Agricultural Marketing Enterprise (AGRICOM) to provide marketing services. - Prices administered on a pan-territorial basis, maize production increase by 165% between 1986 and 1991.	1994 – AGRICOM abolished, debts written off - Instituto de Cereais de Mozambique (ICM) created and acted as buyer of last resort, farmers paid pan-territorial minimum price, manage strategies stocks. Concerns on operations efficiency
Lesotho	1980/81 - Food self-sufficiency program initiated. Producer price set well above world prices – even substantially higher than prices set in South Africa. - Pan-territorial pricing used.	1996/97 - Initiation of policy reform - deregulation of domestic and international grain trade. - More recently, real producer price declined in line with the declining real price of maize grain imports from South Africa used as a reference for domestic price setting.
Botswana	- Producers prices set on cost plus basis. - Pan-territorial & pan-seasonal pricing.	- Move to border parity pricing.

Sources: Donovan (1996), Jayne and Jones (1997), Makenete *et al* (1997), Ministry of Finance and Development Planning, Bostwana, 1997, World Bank, (1997).

PRICE RESPONSE TO LIBERALIZATION

Data collected on food producer and consumer prices in African countries has become a more formidable task in the 1990s than in the pre-reform period. The earlier food channels established by the marketing boards provided an effective avenue for recording prices and volumes of crops sold. The removal of these boards has made data collection more challenging. Much of the food produced in Africa is sold through informal markets and information on prices and volumes from these outlets is fairly scarce. Data available may not be truly representative of the respective average country prices. This caveat needs to be taken into consideration when interpreting some of the data provided in this section. Trends in maize producer prices will be reviewed by examining changes in the real producer price of maize (maize price/cpi ratio) [see Table A11 and A12 for the relative importance of maize in each country].

Producer's Prices

The evolution of real domestic maize producer prices is presented in Table 5.2. Three periods are examined. The first is pre-reform, 1980-84, the second represents a time when many African countries adopted widespread economic reforms, while by 1990 most of these reforms were complete.

Table 5.2: Real Producer Maize Price Trends.

Country	Real Producer Price index of Maize			Coefficient of variation		
	1980-84	1985-89	1990-95/97	1980-84	1985-89	1990-95/97
Nigeria	100	168	198	15	37	32
Benin	100	120	103	38	17	26
Ghana	100	106	86	32	22	18
Togo	100	112	82	4	18	24
Mozambique	100	110	107	18	14	10
Uganda	100	129	115	66	25	31
Tanzania	100	105	152	15	13	33
Swaziland	100	118	81	4	7	18
Kenya	100	108	103	12	4	14
Botswana	100	101	71	11	14	9
Malawi	100	76	77	16	15	13
Lesotho	100	91	78	12	8	6
South Africa	100	89	68	11	15	13
Zimbabwe	100	85	87	14	12	27

The data on maize prices is only complete up to 1997 for Benin. For Botswana, Ghana, Malawi, Mozambique, Nigeria, South Africa, Swaziland, Tanzania, Zimbabwe the end date was 1996 and for Kenya, Lesotho, Togo, Uganda the end date was 1995. All prices are in local currencies.
Source: FAO, World Bank, IMF.

The evolution of these prices is highly linked to policy change. In the early 1980s, maize prices in many Southern Africa countries were set by the state at levels well above the world price in pursuit of their goal of food self-sufficiency. The phasing out of these controls began in the late 1980's through the early 1990's. The price declines in Malawi, Lesotho, South Africa and Zimbabwe resulted from early reform efforts bringing domestic producers in line with lower border parity prices.

The variability in real prices has also changed over time. Between 1980-84 and 1990-1995/97, the maize price variation in seven of the fourteen countries was reduced. In several countries the cause of large price fluctuations, particularly in Southern African countries, is the effects of the weather.

Indeed in 1983 there was a severe drought resulting in higher variability in the 1980-84 period. Again, in the 1990s the effects of the drought were particularly apparent in Zimbabwe.

Observations from Maize Markets in Africa

A more detailed description of several individual country maize markets will be presented in this section to complement the discussion on price trends.

Maize production in **Nigeria** is small relative to cassava, yams and sorghum but still accounts for about 7 percent of food production. The government has maintained a high level of intervention in the pricing, marketing and importing of food. Between 1986 and 1989, nominal food prices increased by between 200 and 500 percent (Donovan, 1996). This increase was largely the result of a decline in production and a ban on food imports. There has also been massive public sector involvement in schemes such as the National Strategic Grain Reserve and the National Buffer Stock Program. The exchange rate has had a tremendous impact on the domestic price of food.

Currently maize is the most important crop grown in **South Africa.** It represents both the major feed grain and the staple food for the majority of the population. Maize accounts for about 40 percent of the value of crop production and about 15 percent of total agricultural production. It is also a major agricultural export. In the past, maize was marketed through a control board established under the marketing act of 1968. This required strict government regulation. Maize was sold under a single channel fixed price scheme, with the Maize Board functioning as a monopolist. In 1987 maize pricing transferred to a pool-type pricing scheme with a gradual move towards a more market based approach. In the past, prices were often set above market clearing levels and recent changes in the grain industry from a cost-plus pricing policy to a market based system led to substantial declines in the real producer price of maize. The 1990s witnessed the abolishment of the maize board with free trade in maize. Recent market developments have included the sale of a number of agricultural commodities, including maize, through SAFEX (South Africa Futures Exchange).

In **Kenya** maize is the dominant food crop, followed by cassava. Like most African countries the Kenya maize market has a long experience of government intervention, dominated by a Maize Marketing Board. After independence, this Board and the Wheat Board formed the National Cereal Producers Board (NCPB). Five hundred new NCPB buying centers were subsequently created (Donovan, 1996). However, as of 1988, financial pressures led to the restructuring of NCPB with a phased closure of NCPB collection depots and debts were written off. In 1994, NCPB was established as a buyer of last resort. The record of reforms in the 1990s has shown inconsistent commitment in allowing free movement of maize in the country and also on price liberalization and restructuring of the state marketing board. Progress in these areas was reversed in the early 1990s. Since 1994, there has been a renewed commitment towards liberalization with agreements for complete liberalization on maize movements and price controls (Donovan, 1996).

In **Tanzania**, the process of maize market liberalization progressed more rapidly than in other countries. In the early 1980s, state support to smallholders largely collapsed and parallel marketing had come to dominate food marketing. Pan-territorial prices were removed together with the statutory monopoly in the procurement and distribution of maize. At the same time private sector participation in maize marketing was encouraged.

In **Malawi**, maize was always traded privately with the Agricultural and Marketing Development Corporation (ADMARC) commonly using private traders as buying agents. ADMARC's functions were to subsidize inputs, stabilize prices and market the maize collected. In 1985-86, as a result of a financial crisis, ADMARC was unable to ensure these functions. The crisis compelled the government to undertake substantial reforms to improve the efficiency of agricultural marketing. Pan-seasonal and pan-territorial prices were removed resulting in a significant expansion of private maize trade. In 1992, additional restrictions were removed allowing a free movement of maize within the country. Licenses for trade are, however, still required. This move to a more liberalized system further resulted in substantial private sector development and private traders paid producers more than the ADMARC price. The Government still sets a floor price for maize.

Zimbabwe, after many years of state control in the maize market, implemented widespread market reform in the late 1980s and early 1990s. The government removed price controls, reduced expenditures on subsidies, relaxed import licensing and foreign exchange controls. There was also a move towards a market-determined exchange rate, elimination of the maize market parastatal deficit, and a new price stabilization role for the marketing board. Even though 17 additional permanent buying stations were established during 1985-92, the number of seasonal rural buying stations declined from 135 in 1985 to 42 in 1989 to nine in 1991 (Jayne *et al*, 1994). This decline suggests that the rapid and extensive investment in rural areas after independence proved to be unsustainable. Government continues to intervene in the maize market particularly in drought years. After the 1992-93 drought, the Grain Marketing Board announced substantially higher official producer prices fixed well in advance of planting. These improved incentives resulted in a bumper crop, while the GMB accumulated unsalable stocks and the trading deficit rose to 2.8 percent of GNP in eight months (Jayne and Jones, 1997). In 1995-96, the Board set a producer price too low relative to market prices and lost market share to private traders and millers. The government continues to set ceiling prices on maize meal.

Maize is the dominant food crop in **Lesotho** accounting for 77 percent of all food produced in the country. Other important food crops are sorghum and wheat. There has been continued government support for the maize producer price with strong intervention in the marketing system. Producer prices were set by the Ministry of Agriculture. Moreover under the Food Self-Sufficiency Program initiated in 1980/81, these prices were set well above world prices and were substantially higher than the prices set in South Africa. Imports, including food aid, constitute over 50 percent of total annual consumption. Most households in Lesotho are deficit producers and formal maize marketing is very limited with only three mills in Lesotho (Lesotho Flour Mills, Maputsoe and Maseru Roller Mills with the later two being owned by Lesotho milling company)[Makente *et al*, 1997]. Most of the maize processed by these mills is from commercial imports. Additionally, due to the high transaction costs, relatively high local prices and inconsistent quality compared to that of maize from South Africa, these companies are reluctant to expand purchases of domestic maize. Import permits for commercial and non-commercial uses of maize are only legally issued by the Lesotho Government, through the Department of Economics and Marketing. Uniform or pan-territorial pricing strategies have been used for the whole country and over the last ten years real domestic producer prices in Lesotho have declined steadily. This decline is due to the falling real prices of maize grain imports from South Africa, which was used as a reference for domestic price setting.

The analysis suggests that the reforms have not resulted in real producer price increases for maize which have consequently had a negative impact of production. While maize producer's do not seem to have benefited much from these market reforms, consumers have received substantial benefits (Box 5.2 as an example).

Box 5.2: Producer Price Declines and Consumer Price Benefits in South Africa.

Agricultural policies in South Africa have long been characterized by pronounced biases towards large-scale farmers. As a consequence, income and wealth distribution, especially in the rural areas, have been skewed in favor of large-scale, overwhelmingly white farmers, most noticeably through land ownership rights (Kirsten and van Zyl, 1996, Deininger and Binswanger, 1995). These policies have included interventions in input and output markets, income transfers and legal restrictions on land ownership. The input policies included subsidies on inputs, tax concessions and cheap credit, while a system of marketing boards provided producer prices that where higher than the corresponding border parity prices for most agricultural products. According to Kirsten and van Zyl (Table 9.10, p 225, 1996) *price support instruments were equivalent to approximately 6 per cent of agricultural revenue* while non-price support transfers to agriculture were equivalent to approximately 10 per cent of the value of agricultural production. These estimates are based on the levels of market price support and non-price support between 1988 and 1993. In recent years most input subsidies have been removed, and as of 1987 price controls on commodities have been reduced as part of a progressive move to more market-based prices.

Townsend and McDonald (1998) used a social accounting based model derived from the price dual (Roland-Holst and Sancho, 1995) to explore the effects of changes in South African economic policies from those in place in 1988. These experiments include examining the income distribution effects of a *6 per cent reduction in agricultural price support*.

The results below, show the consumer price effects of the decline in producer price support. As income increases within racial groups, (income quintiles) the effect, in terms of the reduction in consumer prices, appears to decline. Hence, as would be expected, the poorer groups in South Africa will tend to benefit most from agricultural price liberalization. It is also apparent that Coloured and Black households benefit most and White households least. The clear exception is the lowest income Black households. These households seem to benefit less than the next quintile; this may be a reflection of their relative lack of involvement in market activities due to very low incomes.

Household Effects of a 6 per cent Reduction in Price Support (% change in consumer prices/costs)

Income Quintiles*	Whites	Coloureds	Asians	Blacks*
Lowest Q1	-1.23	-1.77	-1.54	-1.40
Q2	-1.07	-1.75	-1.44	-1.67
Q3	-0.98	-1.68	-1.31	-1.57
Q4	-0.93	-1.42	-1.16	-1.46
Q51 – (ninth decile)	-0.84	-1.30	-1.07	-1.29
Highest Q52 – (upper most decile)	-0.78	-1.14	-0.96	-1.18

* These are the groupings used in the national accounts

Source: Townsend and McDonald (1998)

Consumer's Prices

In a recent study, Jayne *et al* (1995) examined the impact of reforms on the real food price trends in several African countries (Table 5.3). Their report shows that grain and grain meal prices have declined since the adoption of maize market reforms in five of six countries examined. The exception was Zimbabwe where prices increased with the removal of maize meal subsidies keeping consumer prices low. Jayne *et al* (1995) identify the main factors influencing the decline in prices as being: i) better transmission of declining real world prices into the domestic economies by removal of trade barriers (Mali, Ghana); ii) increased food aid in the reform period (Mali and Ethiopia); and iii) increased competition and lower costs in food marketing and processing which reduces marketing margins (Zambia, Zimbabwe, Mali and Kenya). As the major aspects of reform were initiated, miller-to –retail marketing margins appear to have fallen.

Table 5.3: Changes in Retail Food Prices since Pre-Reform Periods.

Country	Crop	Pre-reform Phase 1	% change since pre-reform Phase 2	% change since pre-reform Phase 3
Mali	Sorghum, Bamako retail	100	+16	-21
	Rice, Bamako retail	100	-1	-16
Ghana	Maize, wholesale, average of 3 markets	100	-16	-29
	Sorghum, wholesale, average of 3 markets	100	-18	-38
	Millet, wholesale, average of 3 markets	100	+3	-21
	Yams, wholesale, average of 3 markets	100	+26	+4
	Cassava, wholesale, average of 3 markets	100	+33	-7
Ethiopia	Teff white, Addis Ababa, retail	100	-	-17
	Maize, Addis Ababa	100	-	-11
	Wheat white, Addis Ababa, retail	100	-	-3
	Barley white, Addis Ababa, retail	100	-	-6
Kenya	Official ex-depot maize grain, Nairobi	100	-20	-17
	Official producer price, Kakamega	100	-5	-7
	Refined meal, official retail, Nairobi	100	-5	-2
	Refined Meal (retail plus subsidies), Nairobi	100	-14	
	Maize grain, retail, Nairobi markets	100	-12	-29
Zambia	Official ex-depot maize grain, Lusaka	100	-30	
	Official producer price	100	-26	
	Roller meal, official retail, Lusaka	100	-21	-4
	Roller meal (retail plus consumer subsidy), Lusaka	100	-10	-31
Zimbabwe	Official ex-depot maize grain	100	-29	+21
	Official producer price	100	-16	+24
	Roller meal, official retail	100	+16	+54
	Roller meal, official retail plus subsidies	100	+24	+26
Data based of the following periods.		Pre-reform Phase 1	Phase 2	Phase 3
Mali		1970.10 - 1981.09	10/1981 - 09/1985	10/1985 - 12/1994
Ghana		1980.01 - 1983.09	10/1983 - 08/1985	09/1985 - 12/1990
Ethiopia		1980.01 - 1990.05	-	06/1990 - 12/1994
Kenya		1980.01– 1988.06	07/1988 – 12/1993	01/1994 - 09/1995
Zambia		1980.04– 1986.03	04/1986 – 03/1993	04/1993 - 08/1995
Zimbabwe		1980.04- 1991.05	06/1991 – 05/1993	06/1993 - 09/1995

Source: Jayne *et al, 1995.*

Jayne *et al* (1995) conclude that "The weight of the evidence indicates that consumers, especially urban consumers, have in most cases benefited from the food marketing and pricing reforms initiated in the countries examined (pg.2)". The opening up of the market to private traders has mitigated some of the adverse effects of declining subsidies (transportation, pricing) while lower grain processing costs have reduced the wedge between producer prices and consumer prices (Jayne and Jones, 1997).

Market Integration

The aggregate price trends presented in the previous sections conceal the price variation across grain crop markets within a particular country. In many African countries these price variations can be extreme. In some areas, markets are isolated and food prices tend to only reflect supply and demand conditions in that particular area. The price elasticity of market demand is usually lower in these isolated markets than in highly integrated markets as the confined geographic coverage and lack of sectoral diversity offer limited substitution possibilities (Fafchamps, 1992). As food markets become more integrated, the variance in prices is likely to decline and the yield variability over large geographic markets is likely to smooth out local price disturbances. Benefits not only come from geographic or horizontal integration but also from vertical integration (Table 5.4).

Table 5.4: The Advantages of Market Integration.

Market Integration	
Vertical Integration	**Horizontal Integration**
• Production/logistical economies	• Higher price elasticity of market demand
• Transaction cost economies	• Lower variance of prices
• Risk-bearing advantages	• Reduces correlation between crop revenues
• Advantages in the presence of market imperfections	• Decreases small farmer need to rely on their own food production

Source: Adapted from Jaffee (1995) and Fafchamps (1992).

There is evidence that market integration has improved with the recent changes in marketing policies (see also Box 5.1) (Goletti and Babu, 1994, Dercon, 1995 and Badiane, 1997). Goletti and Babu (1994), in their study of the effects of maize market liberalization on market integration in Malawi, find that the transmission of price signals among various regions of the country has improved as the process of price formation was progressively transferred into the hands of the private sector. The simple idea is that increased spatial price arbitrage would smooth prices between markets and improve market integration. In many African countries, continued investments in marketing infrastructure are required to enhance this effect and Malawi is no exception (Golletti and Babu, 1994). Badiane *et al* (1997) corroborate these results, in a recent study on maize market integration in five African countries. They find that in three of these countries (Benin, Ghana and Malawi) markets have become more integrated. Again in Ethiopia, a similar result was found in the teff markets. Liberalization, and to a lesser extent peace, improved the functioning of the markets and increased the prices paid for teff in the main producing areas (Dercon, 1995). The past teff quota system in Ethiopia suppressed private trade resulting in high spatial margins which have subsequently been reduced. The re-emergence of an effective number of grain traders in African markets is indeed a critical issue. As a large proportion of rural households are actually net buyers of grain, even in normal years, there needs to be a continual encouragement of the emerging private sector as food security will depend on their ability to reduce the costs of food to the grain deficit and generally poor areas.

ENHANCING PRIVATE TRADER EMERGENCE

In a recent study of food markets in Benin, Senegal, Ghana, Malawi and Madagascar, Badiane *et al* (1997) find that the investment response by private traders is largely determined by market location, area covered by traders, liberalization of marketing policies, favorable exchange rate policies, the profitability of trading activity and the education level and experience of the traders. Accelerating the speed of investment response in these markets would require an increase in the supply of infrastructure in the immediate marketing area. This should allow greater opportunities to expand trading activities to cover wider areas and provide a stimulus for higher investment levels.

Mobility Barriers

Many studies on private trader's response and activity (Beynon *et al*, 1992; Duncan and Jones, 1993) indicate a fairly high level of competition and show limited evidence of barriers to entry into primary grain marketing activities. However, when the full spectrum of marketing activities is examined (both primary and secondary), several studies have shown mobility barriers to be a significant constraint to market entry into several marketing niches.

Mobility barriers continue to inhibit widespread private sector entry into all niches across the marketing chain. The extent of these barriers varies at different stages of the chain. Several of them have been identified in a recent study on trader entry into food markets in Madagascar (Barrett, 1997). Subgroups of the marketing chain which require substantial start-up costs or inventories, including activities like wholesale collecting, long-haul transport and inter-seasonal storage, result in relatively closed niches (Figure 5.1). Limited capital access can be a serious barriers to entry. Political risk or credibility of reforms also hinders private entrepreneur's entry into activities with large political uncertainty (ie: storage activities in Madagascar food markets). Access to spare parts and energy supplier networks also provide a significant barrier. In many countries the natural economies of scale in long-haul motorized transport are accentuated by poor road conditions and limited spare parts and fuel availability. Typically, motorized transporters own more than one vehicle in order to realize economies of scale in spare parts inventories and to redirect vehicles so as to ensure cargoes do not perish and important pickup and delivery dates are not missed. The need for a reliable power source seems to have concentrated grain mills near large towns where electricity and fuel stations are available. Mills in the smaller towns uniformly rely on diesel, but electricity is the cheapest and normally preferred source of power in many countries.

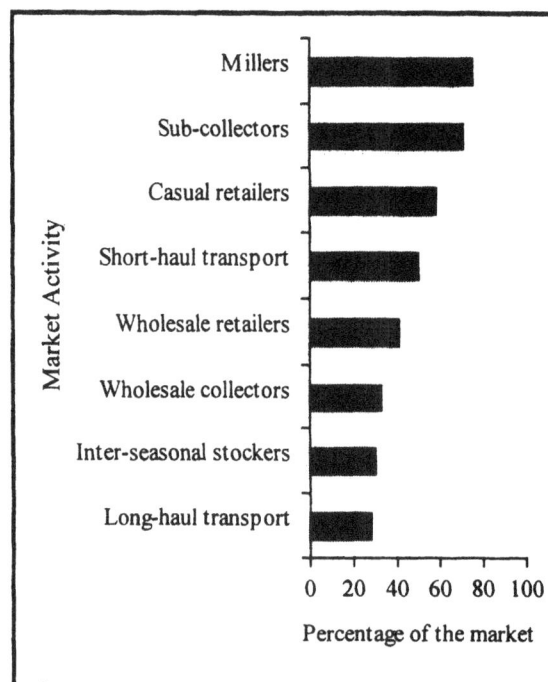

Figure 5.1: Post Liberalization Market Entry into Madagascar Food Markets.
Source: Adapted from Barrett (1997).

86

Box 5.3: Trader Entry and Mobility Constraints in Liberalized Food Markets in Madagascar.

Barrrett (1997) examines trader entry into different niches of the marketing chain in the Vakinankaratra food markets in Madagascar. Liberalization of markets was initiated in the early 1980 and as of 1985 there has been:

- Large private sector entry into markets (52% of sample in Barrett, 1997).
- A significant increase in food prices (60% for rice).

There has been striking intersectoral variation in the proportion of intermediaries who entered since liberalization.

Ranking of the five most important determinants of success by market activity.

	Collectors		Transporters		Millers	Inter-seasonal stockers	Retailers	
	Sub	Wholesale	Short-haul	Long-haul			Casual	Wholesale
Personal rep & relationships	*****	**	*****	**	**	***	****	****
Access to credit	****	*****				*****	**	*
Price paid	***	*	***		*		*****	*****
Ability to evacuate crop	**	***						
Willingness to extend credit	*	****					*	**
Communications link			****	*				
Capacity			**	***	***	***		
Network breadth			*	****		**		
Access to spare parts			*****	*****				
Technology used					***	*		
Supplier network							***	***
Post-liberalization market entry	71%	33%	50%	28%	75%	30%	58%	41%
Mean equipment sunk costs (FMG mn)	0.4	10.5	0.5	197.1	34.3	56.6	0.02	1.9

Source: Adapted from Barrett (1997). Five stars means most important.

Barrett (1997) identifies five main interrelated barriers to intergroup movement within the marketing channel, interlinked contracts, capital access, sunk costs, political risk, access to spare parts and energy supplier networks.

Interlinked contracts - Traders, especially sub-traders (crop collectors) depend on personal relationships for success, a factor which impedes a prospective entrants' ability to separate contracts according to comparative advantage. ***Capital access*** - This can be a serious barrier to entry into subgroups which require substantial start-up costs or inventories. Barrett's (1997) survey results show that 47 percent of the intermediaries in relatively closed niches (wholesale collecting, longhaul transport, interseasonal storage) used bank credit, while for the other sub-groups only 7 percent did. ***Sunk costs*** - The Table above shows the different sunk costs in the market niches, these are negatively correlated with trader entry. ***Political risk*** - Memories of government expropriation of private marketing intermediaries such as silos, vehicles, and warehouses discourages private entrepreneurs from investing in these assets given the continued political uncertainty. ***Access to spare parts*** - natural economies of scale in long-haul, motorized transport are accentuated by poor road conditions and limited spare parts and access to fuel. Most motorized transporters own more than one truck in order to realize economies of scale in spare parts inventory and to be able to redirect vehicles so as to ensure that cargoes do not perish and important pickup and delivery dates are not missed. ***Access to energy supplies*** -Mills are located near large towns as they need a reliable source of power. Moreover, electricity and fuel stations are only available in or near these centres. Mills in the smaller towns uniformly rely on diesel, but electricity is the cheapest and preferred source of power. ***Price*** - An interesting observation from the above table is that price is only viewed as an important determinant for success in the markets with more participants – ie: the more competitive markets. In the case of long-haul transport and storage, price does not seem to be an important issue. Removing these barriers to entry and increasing competition in these niches may improve market efficiency and reduce food prices.

Source: Barrett (1997).

Removing barriers to entry into these marketing niches may improve market integration and efficiency . The government can play a role in removing some of these barriers and in stimulating the confidence and growth of the private sector through the provision of a number of public goods and services. These include stable and transparent policies to improve its credibility, an effective legal system to guarantee property rights and contract enforcement and the development of transportation and communications infrastructure. The latter of these measures is critical. Even though traders may be in favor of reforms and have access to credit, they may still lack incentives to invest in expanding marketing activities due to the high transaction costs created by the segmentation and thinness of the local markets.

Transportation Costs

As highlighted in the previous sections, improving the efficiency of marketing and transport systems can contribute to enhancing producer price incentives and ultimately complement and strengthen price liberalization. An important source of data is provided in an early comparative study by Ahmed and Rustagi (1987), which contrasts marketing costs and price incentives in African and Asian countries. Their study found that average producer prices expressed as a percentage of final consumer prices in the African countries ranged from 30-60 percent, while for Asian countries they ranged from 75-90 percent. Taxes and trade profits justified roughly a third of the differences in price spreads between the selected countries of Africa and Asia, while over two-thirds of the difference was explained by transport and transaction costs. Hence, the level of infrastructure development has suppressed producer prices and inflated consumer prices in African countries compared to their Asian counterparts. In the African countries analyzed by Ahmed and Rutsagi (1987), transport costs alone accounted for 35-40 percent of the total marketing costs of food grains.

The pursuit of policies, such as pan-territorial pricing, has continued to inhibit private sector investment in transportation and storage. In several African countries, other barriers include trade regimes which require licenses for grain trade, and whose effect has been to hinder private imports and exports thus reducing the potential to stabilize food supplies. Unofficial road blocks erected by local authorities to collect money are also common (Niger and Madagascar). Transportation costs vary significantly across African countries with road and transportation infrastructure tending to be better and more developed in the West African countries. Badiane *et al* (1997) show transport costs in Malawi to be about ten times higher than those of some West African countries (Figure 5.2).

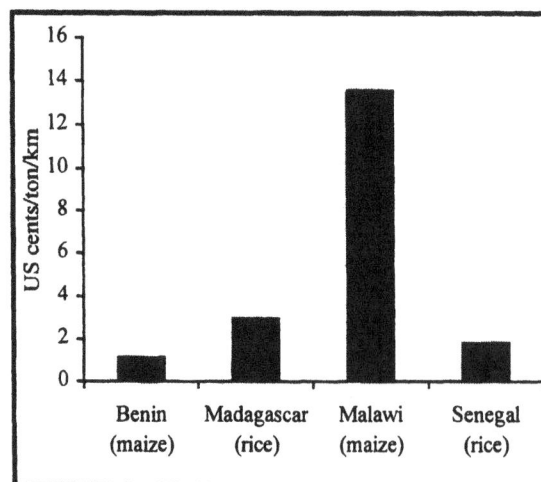

Figure 5.2: Transportation Costs in Four African Countries.
Source: Badiane *et al* , 1997.

While a reduction in transportation costs is likely to raise spatial arbitrage and market integration, higher storage activity is required to increase temporal arbitrage (to smooth out seasonal price fluctuations). Many traders in African countries tend to share storage facilities and, with the uncertainty of the maize market (many substitutes), traders do not store large quantities of cereals for long periods of time. Moreover, lack of familiarity with storage techniques aimed at reducing losses plays against such an activity. Access to credit is also a problem due to high interest rates, lack of collateral, short period of loan term and complicated administrative procedures (Badiane *et al*, 1997).

INCENTIVE POLICIES AND FOOD SECURITY

The challenge of food security is complex due to the plurality of elements that come into play such as supply and demand issues and technology and policy, all of which interact with the three main food security components, namely food availability, food access and food utilization (McCalla, 1998). As there has been recognition that food security is no longer synonymous with food self sufficiency there has been a gradual shift from focusing of food production towards a focus on trade and income generation, issues which will become increasingly important. Although far from exhaustive, this section will briefly elaborate on several policy issues related to food security.

Food security remains high on the agenda of African governments and, as a result, there has been continued interventions in food markets. As a consequence of the need to get food into markets for sale at prices that consumers can afford, governments have been reluctant to divest public storage and processing. However, in many Sub-Saharan African countries, consumer prices have been successfully stabilized through the development of informal marketing channels for low cost commodities. As a large proportion of the rural households are made up, in normal years, of net buyers of grain, the costs of food supplied to the grain deficit and generally poorer areas are critically important to food security. In these circumstances, emerging private trader's ability to reduce these costs will determine the effect of market reforms on these households. Concerns over these food delivery systems were illustrated by the difficulties encountered in marketing and delivering grain from domestic strategic grain reserves and from emergency donor supplies in many of the Eastern and Southern African countries during the severe 1991/92 drought.

Policy linkages with food security are extensive and include employment policy, land distribution, macro-economic growth, distribution of the gains from growth, population growth and income stabilization. A more direct impact on the ability of low-income households to acquire entitlements to adequate food include policies on food production, marketed food supplies, food trade, and food price stabilization. Shifts in domestic food production, for example, directly affect the factor income of household members working in the agricultural sector, whilst changes in food import prices affect the real incomes of households who are net food purchasers. Some of these policy interactions are illustrated in the Mozambique example (Box 5.4).

One often overlooked and critical issues in the policy dialogue on food security is related to gender. Blackden and Bhanu (1998) show through an extensive survey of micro-level studies, that women account for far more of the agricultural labor force than men and they occupy a central position in economic production. Their survey also suggests that women account for more of the time spent on food production than men. However, due to the multitude of tasks women perform in the household (collecting fuel wood and water, preparing meals, etc.) labor constraints severely restrict their response to economic incentives. The labor time allocated to agricultural production is thus primarily spent on ensuring food security. Policies focused on reducing food insecurity in rural households must thus focus on reducing the time burden for women farmers. This includes developing appropriate labor saving technologies explicitly accessible to women covering the full range of their domestic and economic tasks (Blackden and Bhanu, 1998).

Box 5.4: Food Security in Mozambique.

Characteristics

Progress towards improved and sustainable food security in Mozambique has been characterized by:

- increasing per capita calorie availability in the face of dramatic reductions in food aid;
- further lowering and stabilizing prices for the principal domestically produced staple, white maize. An example of this is the average real price of white maize in Moputo which declined by 40%, while the standard deviation declined by 44% between March 1990 and March 1992, and March 1993 to January 1996, pre- and post-drought periods;
- developing a food system now providing consumers with a broader range of low-cost staples from which to choose (Tschirley and Weber, 1996).

Determinants

- the ending of the war
- monetization of yellow maize food aid
- key policy changes:
⇒ removal of movement restrictions across districts and provincial boundaries;
⇒ elimination of the system of official geographic monopolies for registered private traders, which created a surge of private traders;
⇒ market-oriented means of distributing monetized food aid where grain was sold by donors directly to private wholesalers - fueled growth in trading sector and the small-scale maize milling sector.

Mechanisms

- Rural and urban areas have been linked through trade flows - Integration of maize markets between southern, central and northern maize markets has improved dramatically and incentives to producers have increased. As a result, production has increased greatly.
- Maize has been channeled through the small-scale milling sector - over one thousand hammer mills have spread throughout the country.
- Cross-border trade in food products has greatly facilitated changing money in the informal foreign exchange markets.

Future Challenges

Continuing progress towards sustainable food security will depend on:

- consolidating reforms in the trading sector. Food security in the drought-prone southern areas and production incentives in the more productive northern areas will both depend on trade. Simplifying international trade policy and clarifying the national regulatory environment are both important steps of ensuring and strengthening regional and internal trade links.
- investing in cost-reducing marketing infrastructure.
- investing in the country's ability to identify and disseminate improved production technologies as increasing agricultural productivity is essential for long term growth. This will require investments into the research and extension system and private sector input distribution.
- continuing investment to improve the information base on food production, marketing, prices and consumption, as well as on socio-economic characteristics of smallholder households.

Source: Tschirley and Weber (1996).

SUMMARY AND CONCLUSIONS

Market liberalization has been widespread across food markets in Africa. Interventions in these markets tend to have been more extreme in Eastern and Southern African countries.

- Throughout the 1980s, official maize producer prices exceeded export parity prices in the major production regions (Kenya, South Africa, Zimbabwe - typically exporters). These prices have declined to lower border parity levels with market liberalization. In some African countries the decline in real producer prices has been partially offset by improved market efficiency with increased private sector entry into these markets. Between 1985-89 and 1990-95/97, 10 countries of the 14 countries analyzed experienced a decline in the real producer price of maize.

- Lower grain processing costs, made possible through liberalization, have reduced the wedge between producer and consumer prices in many countries. There are several examples where consumer prices have been successfully stabilized in the Eastern and Southern African region through the development of informal marketing channels. Consumers have been the major beneficiaries of the reforms with market liberalization reducing marketing and processing costs.

- In the period since food market reforms, food prices, including maize, have generally declined. Five of the six countries reviewed experienced a decline in real food prices since pre-reforms. Many of these countries have witnessed an increase in competition and lower costs in food marketing and processing thus reducing marketing margins.

- There is some evidence that market integration has improved in many African countries (Ethiopia, Malawi, Ghana and Benin). However, many markets remain isolated with prices reflecting only local supply and demand conditions.

- As a large proportion of rural households are net buyers of grain in normal years, the effects of reform on these households will depend on the ability of emerging private traders to reduce the costs of food to the grain deficit and generally poorer areas.

- Women account for a larger proportion of the time spent on food production than men. Given the multitude of tasks women undertake in the household (collecting firewood and water, cooking, etc.) and the significant labor constraint they face, policies to ensure food security will require the development of appropriate labor saving technologies which cover the full range of their domestic and economic tasks.

- Private traders have emerged to take over the marketing responsibilities in many liberalized food markets in Africa. However, their investment response has been limited where markets are isolated and where there have been backsliding of marketing policies, unfavorable exchange rate policies and low profitability of trading activities. Mobility barriers continue to restrict widespread market entry into several marketing niches across the marketing chain. There has been lower market entry into subgroups of the marketing chain which requires substantial start-up costs.

- High transportation costs in Africa have inflated consumer prices and suppressed producer prices and account for a large share of marketing costs. These costs vary significantly across African countries with road and market infrastructure tending to be better and more developed in West African countries. The pan-pricing structures pursued by many African governments in the past have depressed private investment in transportation and storage.

- Trade restrictions on private imports and exports of grain have impeded the potential to stabilize food supplies and prices.

In many African countries, the private sector has performed relatively well in the liberalized markets with the reduction in the wedge between producer and consumer prices. However, governments have been reluctant to divest public storage and processing due to the food security concerns in many countries. Policy reforms have also been reversed in several African countries with the reintroduction of fixed pricing. If the private sector is to be encouraged to enter these markets, then government commitment to reforms is essential. Other complementary activities again include investing in rural infrastructure, removing trader mobility barriers (political uncertainties, adequate energy supply network and improving access to credit) and removing trade restriction on private grain imports and exports.

6.   FERTILIZER POLICIES, PRICES AND MARKETS

FERTILIZER CONSUMPTION TRENDS

The low fertilizer consumption in Sub-Saharan Africa continues to raise concerns about the continent's ability to overcome its food production problems, which are exacerbated by high population growth rates across the continent[19]. In 1994, fertilizer application rates in SSA were, on average, one fifth of the rates used by other developing countries (Table A16). A large part of this difference is due to the dominance of rainfed agriculture in Africa, compared to irrigated agriculture in regions such as Asia, as well as to the land ecology which dominates in Africa (Voortman *et al*, 1998). Although these averages mask the great contrasts in fertilizer use across countries, with some (Mauritius and Swaziland) applying fertilizer at the same rate as other developing regions, the fact remains that fertilizer use in Africa is disproportionately low.

These low application rates have severe consequences on the fertility of the soil and the sustainability of agricultural production. Evidence suggests that most farmers are not adequately compensating for soil nutrient loss caused by intensive cultivation practices (Donovan and Casey, 1998). Indeed, the declining soil fertility has been highlighted as a major reason for slow growth in food production in Sub-Saharan Africa (Sánchez *et al*, 1995). In other developing countries the 'fertilizer revolution' (ie: in Indonesia, Brazil and China fertilizer consumption has doubled since 1980) has been an integral part of the 'green revolution', without which, the recent progress in achieving food security worldwide would not have been possible (World Bank, 1996).

Green Revolution technology requires using high yielding crop varieties with reliable water supply in combination with macro-nutrient fertilizer application to which these varieties are responsive. A recent study by Voortman *et al* (1998), compares the land ecology of Africa and Asia, noting that the soils in Africa (derived largely from metamorphic Basement Complex rock, which are different from the alluvial and volcanic soils,[20] prevalent in Asia) is one of the primary causes for the lack of replication of the Green Revolution in Africa. While macronutrients (nitrogen, phosphorous and potassium) are important at replacing nutrients most used by crops[21], they suggest that more attention in Africa be paid to correcting chemical imbalances and deficiencies of essential micro-nutrients.

---

[19] Since 1970, cereal production has increased by 2.5 percent per annum which has been exceeded by the 2.9 percent annual population growth.

[20] These soils are relatively young and have a large mineral reserve. The parent material often consists of layers of different materials (tuff, volcanic ash and lava) that differ in both origin and chemical composition and are likely to provide a broad spectrum of plant micro-nutrients. The essential micro-nutrients are less prevalent in the soils in Africa.

[21] Estimates for 38 countries in SSA suggest that annual loss of nutrients per hectare during the 1980s was 22kg of N, 2.5kg of P, and 15kg of K (Weight and Kelly, 1998).

Between 1950 and 1989, about 50 percent of the increase in world food production was due to increased fertilizer use (World Bank, 1989). In Sub-Saharan Africa most of the growth in agricultural output has been the result of area expansion as opposed to gains in yield. Therefore, yield increases will become more important as land becomes scarce.

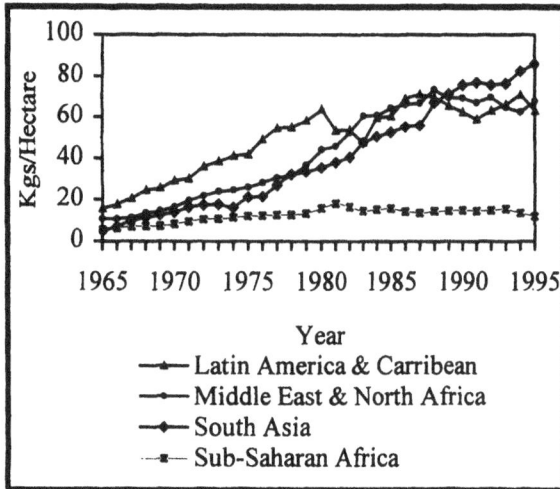

Figure 6.1: Regional Differences in Fertilizer Application Rates.
Source: World Development Indicators, 1998.

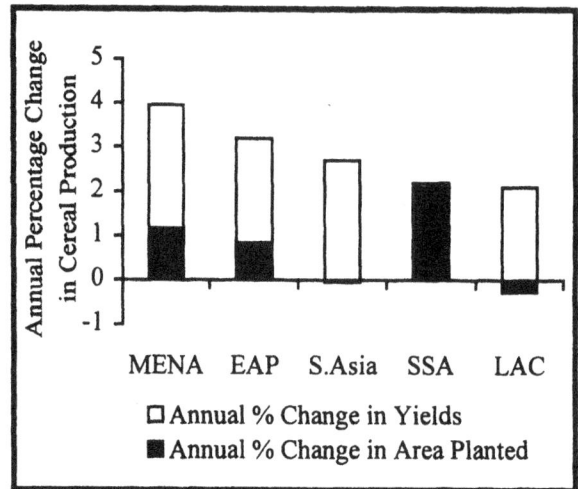

Figure 6.2: Regional Differences in Cereal Production, Area Planted and Yields, 1975-1995.
Source: World Development Indicators, 1998.

THE EVOLUTION OF FERTILIZER POLICIES AND MARKETS

The fertilizer policy environment has had a significant effect on fertilizer markets in SSA. In the late 1970s and the early 1980s, fertilizer subsidies were adopted by almost all countries in the region. This intervention distorted prices and led to an unreliable, high cost marketing and distribution system with a limited choice of basic fertilizers. Explicit fertilizer subsidies were widespread, in 1982/83 prices were subsidized by 25 percent in Malawi, 60 percent in Tanzania, 50 percent in Cameroon, 46 percent in Senegal and 85 percent in Nigeria (Lele *et al*, 1989). Implicit subsidies were also prevalent across almost all countries with an overvaluation of most currencies. It is clear that these subsidies placed a huge direct pressure on government budgets with economy-wide implications. Obviously, removing these interventions clearly goes beyond agriculture[22]. It is apparent that while causing significant budget costs, these subsidies did not significantly effect aggregate fertilizer consumption in Sub-Saharan Africa (Figure 6.1) and thus did not appear to be the most efficient use of public resources to stimulate agricultural production and reduced poverty.

---

[22] Arguments commonly used to justify fertilizer subsidies include: i) compensating farmers for taxes levied through low output prices; ii) trying to overcome farmer risk aversion and help overcome credit constraints; iii) speeding up adoption of agricultural innovations in the early phases of uptake and iii) alleviating the problem of declining soil fertility (fallow systems and other means used to sustain soil productivity in the past breakdown under population pressure). Arguments against fertilizer subsidies include: i) larger farmers being the main beneficiaries; ii) inefficient allocation of resources and iii) public funds devoted to subsidies would be better spent improving infrastructure and investing in other measures aimed at developing an efficient private sector importing and marketing fertilizer. Alternatively expenditures on research and extension would help to spread the availability of more fertilizer-responsive crop technologies and communicate to farmers the best fertilizer recommendations (Donovan, 1996; Lele *et al*, 1989).

Since the late 1980's, with the adoption of structural adjustment programs, many African countries implemented widespread fertilizer sector reforms. Elements of these early reforms were: i) the removal of fertilizer subsidies that reduced the price of fertilizer below border parity levels; ii) public marketing monopolies for import and distribution of fertilizers were retained as they were often seen as more efficient than competitive marketing due to bulk purchasing; iii) the public fertilizer monopoly exclusively handled procurement while attempts were made to liberalize fertilizer marketing at the wholesale level and iv) a large part of the reform effort included privatizing fertilizer retailing.

Recent thinking on fertilizer marketing include: i) encouraging competition and autonomous price formation by supporting appropriate policy and institutions (regulations, commercial and contract law, standard weights and measures, good price information and liberal trade licensing); ii) tariff and non-tariff barriers on all fertilizer imports should be removed and farmers should be given unrestricted access to the full range of fertilizer available on world markets at near world market prices and iii) the private rather than the public sector should be supported by aid-supported fertilizer imports (Gisselquist 1994, Meerman 1996).

By the early 1990s, most countries in SSA had removed their fertilizer subsidies and liberalized their distribution system with the encouragement of private sector participation. In a recent World Bank study of 29 countries, 22 of them had extensive fertilizer market controls and price subsidies in the past. By 1992, 14 of these 22 countries had removed these controls completely (World Bank, 1994). Only two countries, Nigeria and Malawi, retained extensive intervention. In recent years, both countries have liberalized their fertilizer markets. However, there are some concerns about backsliding on previous policy reforms in many countries. Although subsidies have been removed and internal distribution has been liberalized, non-tariff barriers in many countries (Benin and Ghana) continue to inhibit fertilizer trade. Several countries in SSA produce fertilizers (Burkina Faso, Côte d'Ivoire, Mauritius, Nigeria, Senegal, South Africa, Tanzania and Zimbabwe) but the bulk of the continent's fertilizer requirements are imported. Many African countries continue to rely almost exclusively on fertilizer aid.

The removal of fertilizer subsidies has clearly had a beneficial effect on the fiscal balance. In many countries, the subsidy removal and devaluations of the domestic currency have increased fertilizer prices significantly. A general argument is that with the reduction of export taxes and liberalization of agricultural commodity markets, higher fertilizer prices would be offset by higher export crop prices. A common concern in SSA is that export crops are grown by a small number of usually better off farmers while most of the poorer households rely on grain production. Maize accounts for about 25 percent of fertilizer consumption in Sub-Saharan Africa while export crops such as tea and coffee account for less than one percent (Kelly *et al*, 1998). Grain farmers thus carry the full burden of higher fertilizer prices while not enjoying the benefits of higher export crop prices. This 'cost squeeze' has a significant effect on farm income and fertilizer use. There have also been concerns that farmer access to fertilizers has been reduced by removing parastatals that distribute and supply credit for fertilizer purchases. Past experience suggests that input markets serve farmers best when there is some degree of vertical co-ordination among input distribution, output marketing and credit functions, which lowers costs and improves loan repayment rates. The most successful examples of vertical co-ordination have been in sub-sectors producing industrial or export crops (cotton, for example). In such cases, increased access to improved inputs and more reliable output markets stimulate productivity in food as well as in cash crops (Kelly *et al*, 1998).

PRICES AND FERTILIZER CONSUMPTION

Several price and non-price factors have been used to explain fertilizer use in Africa. These are summarized in Table 6.1 and include profitability of fertilizer use, human resources availability, financial liquidity, household assets, market access and extension services. Some of the non-price explanatory variables implicitly impact on price variables eg: the distance from the fertilizer market affects the fertilizer price, highlighting the complementarity between the two.

Table 6.1: Summary of Fertilizer Adoption and Use Intensity Studies in Africa.

Variable Groupings	Explanatory variables Included in the studies	Adoption of fertilizer — Sign	Adoption of fertilizer — Study/Authors	Fertilizer use Intensity — Sign	Fertilizer use Intensity — Study/Authors
Profitability	Fertilizer price			-	Kimuyu *et al* (1991)
	Crop price			+	Kimuyu *et al* (1991)
	Crop-fertilizer response			+	Adugna (1997)
	Improved variety	+	Green and Ng'ong'ola (1993)		
Human Resources	Age	-	Jha and Hojjati (1994)		
	Dependency ratio	-	Adugna (1997)		
	> 5 yrs of schooling			+	Jha and Hojjati (1994)
	Literate household head	+	Demeke *et al* (1998)		
	Female headed households	+	Demeke *et al* (1998)		
	Farm labor	+	Mbata (1997), Green *et al* (1993)		
	Hired labor used	+	Adugna (1997)	+	Adugna (1997)
Financial Liquidity	Farm income	+	Mbata (1997)		
	Off-farm income	+	Adugna (1997)	+	Adugna (1997)
	Total farm sales	+	Jha and Hojjati (1994)		
	Cash crop grown	+	Green *et al* , Demeke *et al* 1998		
	Credit	+	Mbata , Adugna (1997), Green *et al*	+	Kimuyu *et al* (1991), Adugna
Household Assets	Use oxen			+	Jha & Hojjati (1994)
	Number of oxen owned	+	Adugna (1997)	+	Adugna, Demeke *et al* (1998)
	Farm size	+	Mbata (1997)	+ (-)	Demeke *et al* , (Adugna)
	Maize area	+	Mbata (1997)		
Market Access	Distance to fertilizer market	-	Jha & Hojjati; Demeke *et al* 1998		
	Fertilizer transport costs	-	Mbata (1997)		
	Member of Co-operative	+	Jha - Hojjati (1994), Mbata (1997)		
Extension	Extension	+	Mbata , Adugna , Demeke *et al*		

Source: Green and Ng'ong'ola (1993), Kimuyu *et al* (1990), Jha and Hojjati (1994), Demeke *et al* (1998), Mbata (1997), Adugna (1997).

In Africa, many of these non-price factors are critical. Market access is a key constraint to many farmers who have to travel long distances to reach markets. Access seems to improve with co-operatives which can enhance economies of scale in purchasing. Lack of financial liquidity is also a key constraint to fertilizer adoption and the intensity of fertilizer use. In the past, grain farmers had greater access to credit through some of the marketing boards. Farmers, lacking resources and assets with differing attitude towards risk, are less likely to adopt fertilizer. Human resources and extension services are positively correlated to fertilizer adoption. Increased knowledge of improved farming techniques along with availability of resources to apply this knowledge are likely to increase fertilizer use (Demeke *et al*, 1998). Figures 6.3 and 6.4 show the range of values (minimum to maximum) for two incentive ratios that have been observed in Africa.

Although non-price factors are important, prices remain an key determinant of the incentives provided to use fertilizer. There has been very little documented work on estimating demand elasticities for fertilizer in Africa. Most studies remain descriptive and lack empirical content due to insufficient data to allow quantitative analysis. Several studies have attempted to estimate these elasticities for African countries showing that farmers are price responsive. Price elasticities of –0.82 to –1.08 were derived for Malawi smallholders (Chembezi, 1989) and –0.47 for South Africa (Khatri *et al*, 1996).

Profitability remains one of the key factors determining the quantity of fertilizer used. Farmers will not use fertilizer if it is not profitable. Profitability is determined by the price of the crop grown, the price of the fertilizer used and the output response to the fertilizer application. Profit incentives are usually measured by the value/cost ratio[23], with a general consensus that a ratio of greater than two is required for farmers to adopt fertilizer. It has been suggested that in high-risk production environments, the minimum ratio for adoption is three or four (Kelly *et al*, 1998).

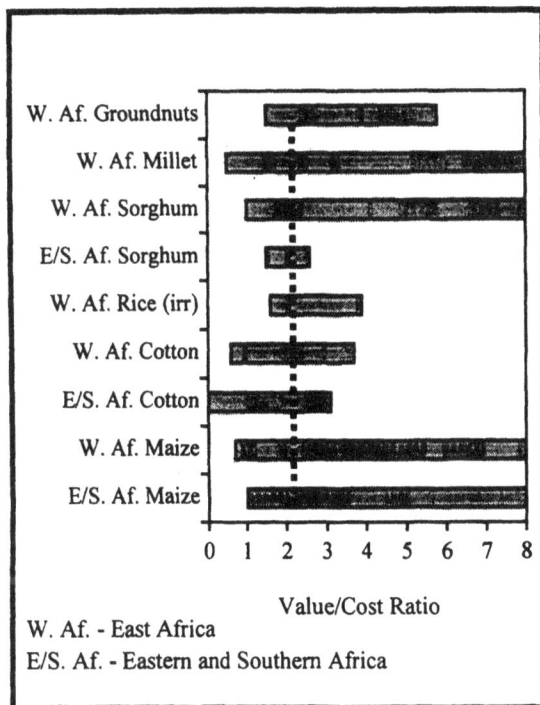

Figure 6.3: Fertilizer Profit Incentives.
Source: Kelly *et al* (1998).

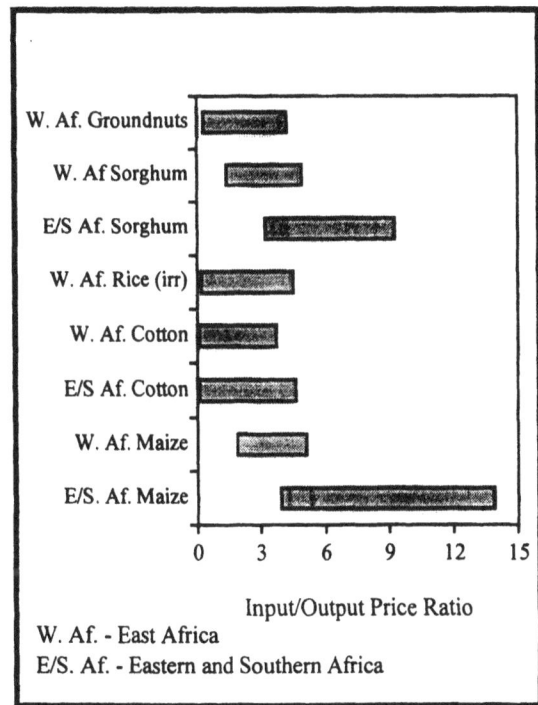

Figure 6.4: Fertilizer Price Incentives.
Source: Kelly *et al* (1998).

Estimates of the value cost ratio suggest that the profitability of several crops in Sub-Saharan Africa are favorable to fertilizer use (Figure 6.3). The output/nutrient ratio is also commonly used to analyze fertilizer incentives. It shows how many kg's of additional output a farmer can obtain from a kg of fertilizer nutrient. This agronomic response varies widely among countries and regions, and although

---

23 The value cost ratio is the product value attributed to fertilizer use divided by the cost of fertilizer. Alternatively it can be defined as the product unit price to fertilizer unit price multiplied by the fertilizer response rate.

useful, it may be quite limiting unless additional information on how it was derived is given. Donovan and Casey (1998) provide a critical review of these technical considerations highlighting constraints to improving soil fertility and recommend actions to address them. The output/input price ratio is also frequently discussed as there is evidence that farmers use it in making their decisions (Kelly *et al*, 1996). These price incentives for several crops are highlighted in Figure 6.4.

THE EVOLUTION OF FERTILIZER PRICES

Enhancing these fertilizer adoption incentives will require improving crop prices relative to fertilizer prices, reducing fertilizer prices relative to crop prices or improving the agronomic response of fertilizer application. This section will focus on explaining the evolution of fertilizer prices and will examine the domestic price - world price ratio, identifying some of the key factors influencing price differentials across countries.

World Fertilizer Prices

Since 1980, real world fertilizer prices have declined by about 40 percent while the fertilizer price/grain price ratio has remained relatively constant. The fertilizer price/export crop price ratio has been fairly unstable but the 1997 ratio was similar to the 1980 ratio (Figure 6.5). The fluctuations have been largely caused by large export crop price changes. The trends in the world prices since 1980 suggest that the price of fertilizers have not significantly increased relative to the world price of exports and grain crops.

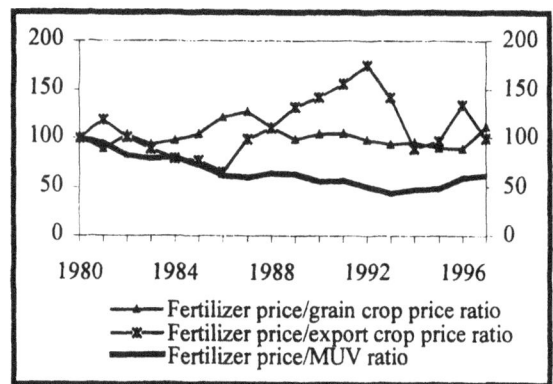

Figure 6.5: World Fertilizer Price Ratios.
Source: World Bank data.

Domestic Prices

In the post liberalization period, real farm-gate fertilizer prices have increased in most countries (Figure 6.6) largely due to the removal of subsidies, the increase in the world fertilizer price and the devaluation of domestic currencies. Urea provides an interesting example[24]; most countries experienced an increase in the real price of urea with more than a 50 percent rise in four countries (Cameroon, Ghana, The Gambia and Malawi [Figure 6.6]).

The results suggest that in most cases, where a comparison can be made between an earlier period and a later period, the price ratio is higher in the later period (Figure 6.7). The changing ratio is

---

[24] Cross-country price data on urea was readily available and was therefore used as an example.

likely to significantly influence fertilizer consumption. These relationships will be examined more closely in the case studies presented in later sections.

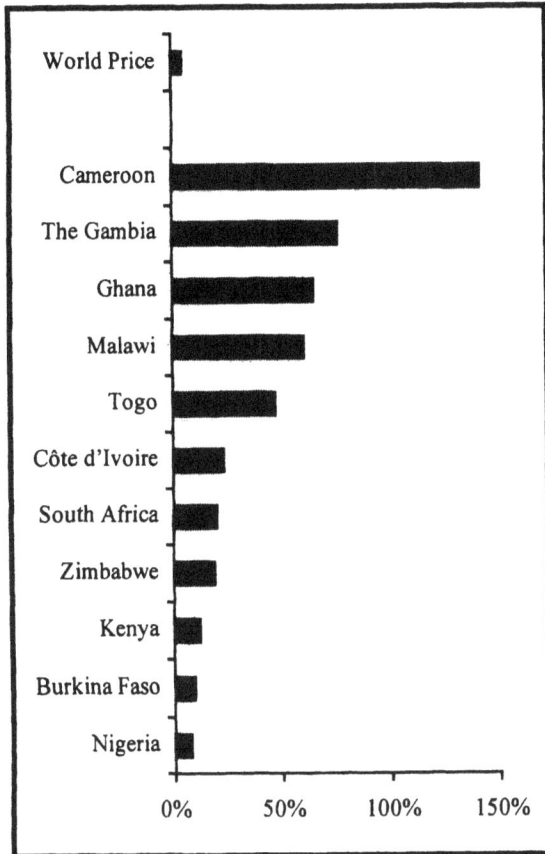

Figure 6.6: Change in Real Fertilizer Prices, 1990-1995/97.
Source: World Bank data.

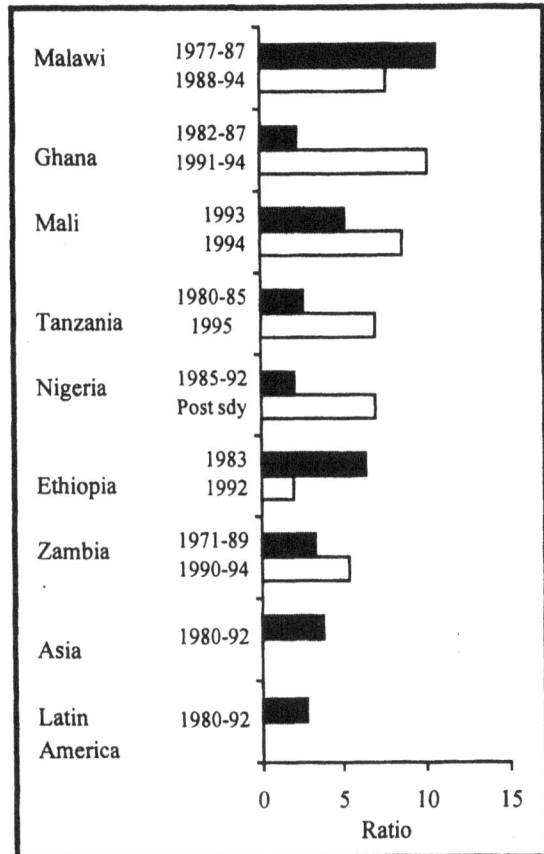

Figure 6.7: Nitrogen/Maize Price Ratios.
Source: Heisey and Mwangi (1996),Gerner et al (1996).

EXCHANGE RATE PASS-THROUGH TO FARMERS

Currency devaluations appear to have been passed onto higher fertilizer prices more effectively than the pass-through to higher export crop prices (examined previously). Table 6.2 summarizes the information contained in Figure 6.8. In Malawi, The Gambia, Nigeria and Cameroon (the countries above the forty five degree line) the price increases are in excess of the devaluations and world price increases. This suggests that in these markets the pass-through of fertilizer costs to farmers has been in excess of the currency devaluation, thus eroding the incentives to use fertilizer[25].

After removing the effects of devaluation and world price increases from the domestic urea price, six of the eleven countries analyzed (the countries below the forty five degree line) have reduced

---

25 For fertilizers, the countries above the 45 degree line were labeled as having inhibiting pass-through while for crops, countries with inhibiting pass-through appear below the 45 degree line. This analysis focuses on the pass-through to farmers. If there is a real currency devaluation of 50 percent and fertilizer prices increase by 70 percent then farmers loose; which was considered as inhibiting. This is converse to export crop pass-through discussed in chapter four.

in Figure 6.8 do not take into account annual price variations. This is clear in the case on Malawi, which experienced huge fertilizer price increases in 1996. An analysis of this annual change would show Malawi way above the line (Figure 6.8).

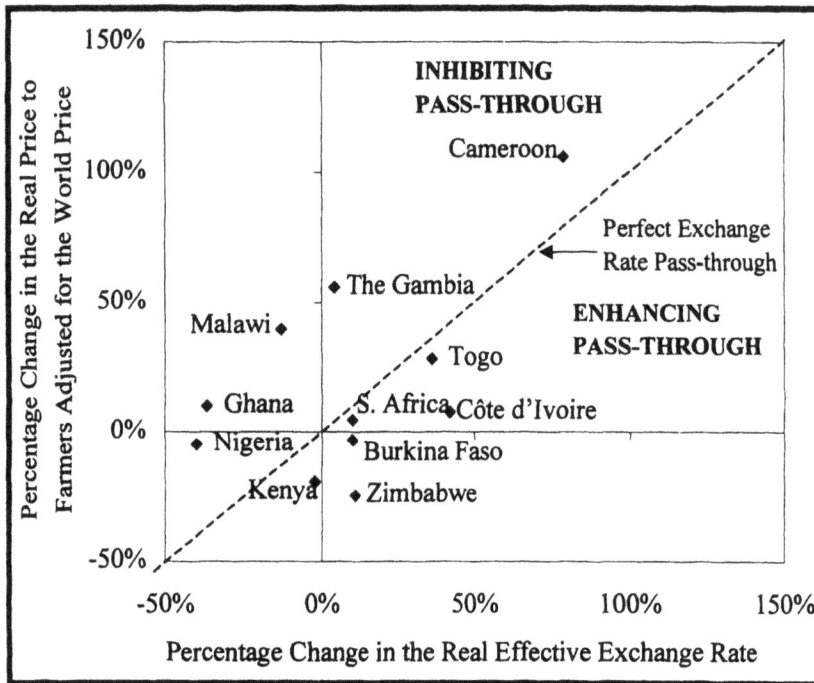

Figure 6.8: Exchange Rate Pass-Through to Fertilizer Prices, 1990-1995/96.

Table 6.2: Difference between the Percentage Change in Real Urea Prices and the Percentage Change in the Real Effective Exchange Rate.

Country	Exchange Rate Pass-Through
Malawi	53
The Gambia	52
Ghana	47
Nigeria	35
Cameroon	27
South Africa	-5
Togo	-8
Burkina Faso	-13
Kenya	-17
Côte d'Ivoire	-34
Zimbabwe	-36

Source: FAO and World Bank data.

Transportation: Most Sub-Saharan African countries import their fertilizer as local production proves less profitable due to the small size of the local markets (Heisey and Mwangi, 1996). Nonetheless, the costs of fertilizer imports are high and the difference between landed costs and the world f.o.b. price tends to be twice as high in many Sub-Saharan countries as they are in Asian countries (Shephard and Coster, 1987). These large differences have been attributed to high transportation costs of the small fertilizer volumes most African countries import (Bumb, 1988) [see also Box 6.1]. A small trading share also weakens the nation's position in negotiating for lower prices. Internal distribution costs also tend to be considerably higher than in other developing countries; this is largely due to poor development and maintenance of the domestic infrastructure (Heisey and Mwangi, 1996). Most fertilizer trading companies and their representative agents rent the services of private transport companies and do not own their own transportation vehicles. Transport costs can sometimes constitute 50 percent of the operating costs of traders. The transportation problems experienced by input traders and highlighted by Badiane *et al* (1997) were due to excessive costs (Benin), transport restrictions (Ghana), frequent road tolls (Madagascar) and poor quality of roads (Malawi). Late delivery is also common resulting from credit requirements throughout the fertilizer marketing channel. These costs persisted in the past as governments were typically slow to relinquish their involvement in fertilizer importing and marketing, as fertilizer was considered a strategic input.

100

Trade restrictions and regulation: Continuing trade controls restrict private sector entry and competition in fertilizer markets, contributing to high margins and prices in many countries. Several countries have a restrictive list of fertilizers that can be imported (Ghana) which denies farmers the opportunity to use nutrient-wise, more cost effective fertilizers. The lack of competition in fertilizer distribution systems and trade, often a result of public sector operations, can also contribute to inefficiencies and higher marketing margins (Pinstrup-Andersen, 1993, Gisselquist, 1994).

Fertilizer regulations have also been restrictive. Farmers have been cut off from fertilizer markets in neighboring countries, forcing importers to consider countries as a separate and small market (Gisselquist, 1998). Private traders in many countries are not able to import fertilizers without prior government approval involving time consuming and costly procedures to satisfy multiple agencies. Packaging regulations also need to be relaxed. Truth-in-labeling needs to be enforced to control fraud and prevent the sale of substances that are harmful to the soil. This involves making rules for sampling and setting ranges of allowed variance around stated compositions for various nutrients (Visker *et al*, 1995).

In many countries, an avenue for reducing these high prices without substantial cost may be through the elimination of trade restrictions. In some instances, as international experience suggests, the negative impact of higher fertilizer prices can be offset to some degree by trade liberalization. As an example, Turkey in the 1980s protected domestic fertilizer producers with import duties and then subsidized fertilizers to farmers, the net impact of these interventions left farmers with near world-market prices, so that cutting import barriers and removing subsidies at the same time could have been expected to leave domestic fertilizer prices essentially the same (Gisselquist, 1996).

Macroeconomic policy: Macroeconomic policies have a significant impact on fertilizer prices. These have included currency overvaluation, budgetary constraints and foreign exchange restrictions all of which have been prevalent in African countries. Overvalued exchange rates reduce fertilizer prices while both budgetary and foreign exchange restrictions raise prices.

Fertilizer aid: Sub-Saharan Africa receives extensive fertilizer aid, which has had an inhibiting effect on domestic fertilizer markets. These aid programs are characterized by lack of competition in procurement and distribution, higher fertilizers costs and slow delivery of inputs. Moreover, many of the programs are vulnerable to mismatching of inputs with needs. These aid programs need to be radically changes or abolished (see later section on fertilizer aid). In the meantime, the beneficiary countries can improve the usefulness of this aid in a number of ways. They can create greater transparency and consistency in preparing the list of aid requirements (fertilizer, chemicals and machinery) that they submit to the donor country. They can also enhance the speed at which relevant documentation is completed so as to ensure that assistance is provided in a timely manner. Recipient countries must ensure that the sale of the aid (fertilizers) is at market prices and that government subsidies on distribution and storage must be eliminated. In some cases, a counter-part fund is required to be set up in the recipient country. The use of the counter part fund needs to be clearly identified and should be consistent with the overall agricultural strategy. The uncertainty of government actions needs to be removed as it could undermine the development of private sector activity in these input markets.

101

Box 6.1: Explaining the Differences in Producer Price/World Price Ratios of Urea Across African Countries.

The high fertilizer prices in Sub-Saharan Africa have raised concerns about the subsequent fertilizer adoption rates (determinants of which included the profit incentive [the value cost ratio], the price incentive [the input-price ratio] and intensity of usage). Awareness and understanding of the factors that cause differences in the domestic price/world price ratio for fertilizer among African countries may provide useful policy information for addressing distortions in these markets. An econometric analysis may provide some insights.

The difference in the domestic price/world price ratio across countries was expressed as

$$\frac{PD}{PW * e} = \alpha + \beta_1 DISTANCE + \beta_2 TARIFFS + \beta_3 ROADS + \beta_4 VOLUME + \beta_5 EXCH + u$$

where PD is the domestic price of fertilizer, PW is the world price, e is the nominal exchange rate, α is a constant, DISTANCE, the distance from international fertilizer markets, TARIFFS, the fertilizer import tariff, ROADS, the percentage of paved roads, VOLUME, the volume of fertilizer consumed and EXCH, the exchange rate policy stance (parallel market premium). Ordinary least squares were used for the estimation and the results are presented in the table below.

	Variable	OLS		OLS with White's Heteroscedasticity Adjusted S.E.'s
		Coeff.	t-stat	t-stat.
	CONSTANT	2.4483	2.4	3.4
Location	DISTANCE	0.0003	3.8	3.8
Transport infrastructure	ROADS	-0.0309	-1.9	-1.8
Trade policy	TARIFFS	0.0713	2.3	2.0
Macroeconomic policy	EXCH	-0.4106	-0.7	-2.4
Volume consumed	VOLUME	-0.00002	-0.9	-2.6
	R^2	0.53		
	Observations	31		
	F-stat (5, 26)	5.69 (0.001)		

The variables included in the model explain only about 50 percent of the cross-country variation in the domestic price/world price fertilizer ratio. The relatively poor fit may be due to omitted variables or simply to the poor nature of fertilizer price data in Africa. The best model in terms of goodness of fit, data coherence, parsimony in parameters and consistency with theory (Hendry and Richard, 1982), is presented in the table. Although the R^2 is low, the t-statistics shows some significant variables. A common phenomenon with regressions using cross-section data is heteroscedastic error terms. Diagnostic test on the error suggests possible heteroscedasticity and this was corrected using White's adjusted variance-covariance matrix (White, 1980).

The *distance from international fertilizer markets* and the *fertilizer import tariff* have positive coefficients while the other variables have negative coefficients. The distance variable has the most significant effect on the price ratio and suggests that the fertilizer price ratio increases with an increase in distance from international markets. Similarly, the higher the *import tariff* the higher the domestic price. These obvious explanatory variables are consistent with a priori expectations. The higher the *percentage of paved roads* in a country, the lower the price ratio, suggesting lower domestic transportation costs. If large *volumes of fertilizer are consumed* by a particular African country, economies of scale in fertilizer trade can be realized, resulting in a lower price ratio. The larger demand will also attract more private traders. The final variable is the exchange rate where *overvalued exchange rates*, subsidize fertilizer prices resulting in a lower price ratio.

The emergence of private traders: Since liberalization of the fertilizer markets in the countries studied, many multi-national companies have entered the market. However, parastatals continue to operate and dominate these markets in many countries. In certain countries, only a few private companies have penetrated these input markets creating an undesirable oligopolistic situation which has had negative effects on prices. The results from a recent IFPRI study suggest that liberalization of input markets in most of the countries studied has been partial. Thus, it has restricted the opportunities for private traders to enter the market and created uncertainty about the change in policy direction[26].

The IFPRI study identifies several factors that continue to inhibit the industry: these include supply problems, lack of effective distribution, timely delivery and transportation. The manufactured products, particularly the compound fertilizers, are of relatively low analysis which increases the cost of handling, transportation and storage per unit of nutrient.

OBSERVATIONS FROM FERTILIZER MARKETS IN SEVERAL SELECTED AFRICAN COUNTRIES

Policy reforms have played an integral part in the changing fertilizer markets in SSA countries. The previous analysis clearly illustrated this and continuing impediments on fertilizer prices were identified in several countries. This section will attempt to expand on some of these constraints by focusing on individual issues through country case studies. Four countries will be examined: Malawi, Zimbabwe, Ghana and Ethiopia.

Addressing the Fertilizer Aid Issue in Africa

Agricultural fertilizer markets in Africa are not performing well. Only seven countries in Sub-Saharan Africa produce fertilizer (Burkina Faso, Mauritius, Nigeria, Senegal, Zambia, Zimbabwe and South Africa). Only four of these (Nigeria, Senegal, Zimbabwe and South Africa) export it. Many African countries rely almost exclusively on foreign aid (Figure 6.9). In 1987, twenty African countries relied almost exclusively on fertilizer aid to meet their domestic requirements (Bumb, 1988). Such dependence has caused uncertainties in planning, procurement and distribution of fertilizer products. Bumb *et al* (1994) and Ndayisenga and Schuh (1995) have underlined these uncertainties.

---

[26] The International Food and Policy Research Institute have recently completed a major study on agricultural input markets in Africa (Badiane *et al*, 1997). Market surveys were conducted in five countries (Benin, Ghana, Madagascar, Malawi and Senegal) to raise the understanding of conditions for the successful reform of agricultural inputs markets in African countries and to identify strategies to complete the transition to competitive and efficient output distribution and input delivery systems. Several of their key findings are highlighted above.

- Supply uncertainties are created as long-term commitments from donors to provide fertilizer are infrequent. Recipient countries are left to depend on the annual decisions of donors.
- Product choice is limited with donors offering countries a list of fertilizers from which to select. This list usually reflects the fertilizers in excess supply in the donor country as opposed to those which are needed by the recipient country. This results in a disconnect between aid fertilizer given and the needs of the recipient country.
- Domestic fertilizer markets in the recipients countries are disrupted. In some cases, the price of aid fertilizer is higher than that of domestic market prices while in other cases the price is highly subsidized. These uncertainties inhibit the development of the private sector.

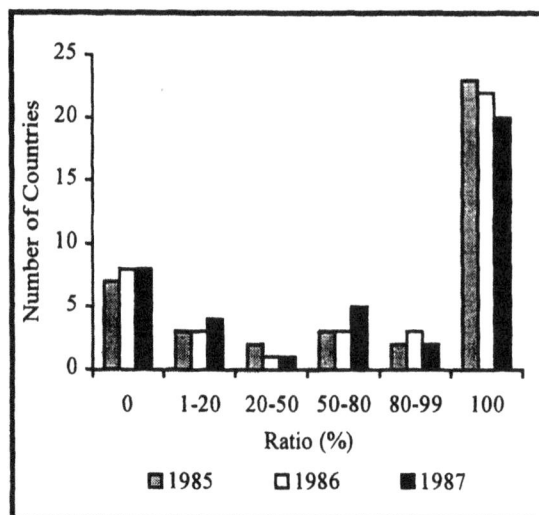

Figure 6.9: Distribution of Sub-Saharan African Countries by the Ratio of Fertilizer Aid to Fertilizer Imports, 1985-87.
Souce: Adapted from Bumb *et al* (1988) citing FERTECON 1989.

- In some cases, fertilizer aid may have adverse long-term macroeconomic implications. In many cases 'free fertilizer' (which is provided by some aid programs) provides significant macroeconomic benefits to recipient countries through savings of foreign currency that would have been used to import fertilizer. If these aid inflows are significant, the domestic currency may strengthen, shifting domestic terms of trade against the tradable sectors - like for agriculture which is counter-productive in terms of providing adequate incentives to agricultural producers.

An Example of Fertilizer Aid: The 2KR Japanese Aid Program to Sub-Saharan Africa[27]

Japan's Grant Aid for the Increase of Food Production is supplied to all regions in the world. The annual budget amounts to US$200-300 million and is allocated to 48 countries (JICA). In 1996, Sub-Saharan Africa received the largest amount of this aid worldwide and accounted for 41% of the total 2KR grant aid. The remaining aid was distributed to Asia (28%), Central and South America (15%), the Middle East (8%) and East Europe (8%).

The magnitude of Japan's generous contributions suggests that it can play a major role in helping improve the fertilizer distribution networks in many Sub-Saharan African countries. Unfortunately, the 2KR program in SSA often works counter the goals of free and sustainable markets for fertilizer. Efforts to encourage the emergence of strong private sector networks are often undermined by the market distortions created by the 2KR process.

[27] This is a brief summary a longer paper titled 'An Overview of the Japanese 2KR Aid Program' prepared by Tatushi Adachi and Robert Townsend. The program is known as 2KR, after the Second Kennedy Round of trade negotiations in 1977. Under a special budgetary provision, fertilizer, chemicals and machinery were supplied by Japan through a new network 'Aid for the Increase of Food Production' which is known as 2KR aid.

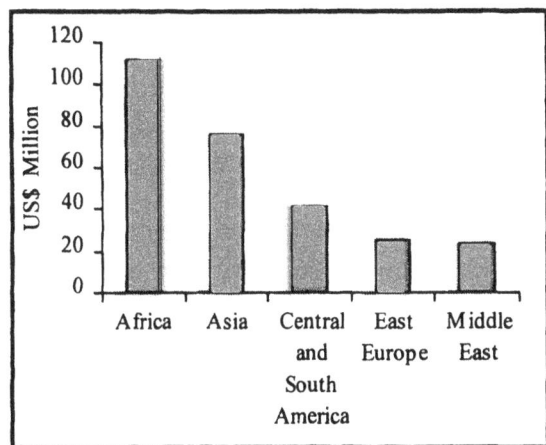

Figure 6.10:Regional Distribution of 2KR Aid, 1996.
Source: JICA.

In 1996, Africa received more 2KR aid than any other region (Figure 6.10) with 26 countries receiving this aid (Figure 6.12). The distribution of aid varies greatly among and within regions. In 1993, SSA received significantly more pesticides than any other region in the world (Figure 6.11). Moreover, the volume of fertilizer aid was second to Asia.

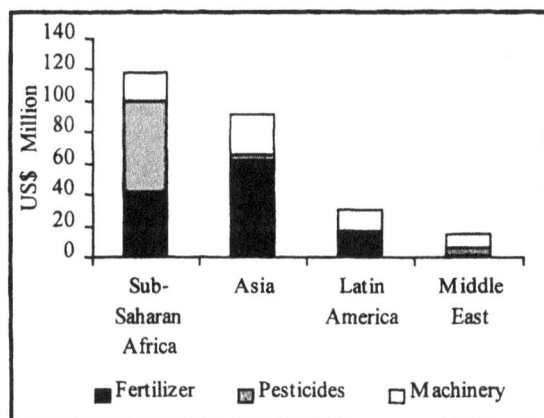

Figure 6.11: Regional Distribution of 2KR Aid, 1993.
Source: Tobin (1996).

The largest African recipients of 2KR include Kenya, Tanzania, Ethiopia and Zambia. Kenya receives one billion Yen worth of agricultural inputs from the 2KR program, which makes it the fourth largest 2KR recipient in the world, after the Philippines (¥1.7Mil), Sri Lanka (¥1.4Mil) and Indonesia (¥1.3Mil).

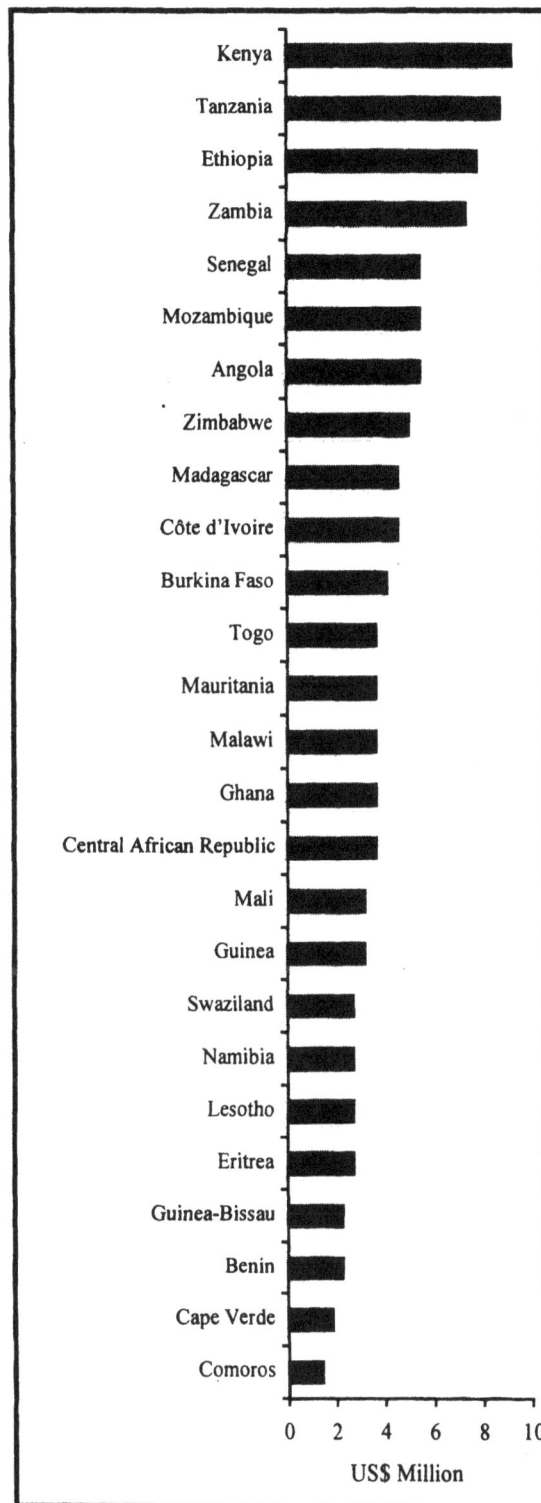

Figure 6.12: Distribution of 2KR Aid by African Country, 1996.
Source: JICA.

105

The process of supplying fertilizers under the 2KR program is not dissimilar to the programs of other donors. Formal requests are made to the government of the donor country, discussions are held to assess the merits of the request and ability of the donor country to supply the goods, the donor opens a restricted tender for the requested goods, an award is made to the lowest bidder and counterpart funds are deposited by the recipient country into a designated domestic account upon sale of the goods.

From the recipient country perspective, fertilizer supplied under the 2KR program has a particularly serious negative impact on private sector participation in fertilizer markets in Sub-Saharan Africa. This is due to the characteristics of the aid program.

- *Lack of competition:* The most apparent trait of the 2KR program is the lack of competition in several stages of the program cycle, most notable in procurement and in the agricultural input markets of the recipient countries.
  Procurement: Restrictions on the number of traders who can participate in the program creates significant rent seeking activity. This appears to have been more prevalent in the 1980s when aid was tied exclusively to Japanese fertilizer and procured through Japanese trading companies. The tendering process appears to be almost un-contested which results in a high f.o.b. (Japan) and c.i.f. price (recipient country) of inputs relative to competitive market costs. Japanese traders who shipped the inputs to Africa appear to have received unprecedented profits.
  Domestic markets: Recipient country governments usually distort the domestic markets of inputs received under the program. Fertilizers are usually sold by the government on local markets at subsided prices (or provided free) which inhibited the development of private sector involvement in fertilizer distribution and storage.
  The current mechanisms of procurement and distribution has resulted in *leakages* from the system in terms of exorbitant intermediary profits. The lack of competition has resulted in a high price of inputs acquired under this scheme. In the 1980s it appeared that many countries experienced difficulties in generating counter part funds which are required under the program because of these high costs.
- *Lack of integration:* The lack of integration of 2KR fertilizers into the domestic fertilizer market discourages private sector investment in fertilizer importation and marketing. The 2KR fertilizers are typically channeled through either a government-owned/controlled entity or through a firm linked to the Japanese supplier of 2KR fertilizer. In several countries, attempts have been made by governments to negotiate with the private sector to distribute the 2KR fertilizer. However, in some cases, the government was unable to recover the price 'paid' which prevented further negotiations with private traders who claimed to be able to acquire fertilizers form other suppliers at a price lower than the price paid under the 2KR scheme. In this case, the private sector was able to import the inputs at lower cost. In some countries, the recent untying of aid has lowered the price of fertilizer and pesticides significantly.
- *Slow delivery:* The delivery of 2KR aid to the recipient countries is notorious for its delay which has a significant impact on the effectiveness of the program. Late deliveries of fertilizer prove ineffective for farmers and create additional costs through storage until the next cropping season. The primary source of delay is the complicated diplomatic and bureaucratic process that is involved in 2KR procurement.
- *Vulnerable to mismatching of inputs with needs:* There also seems to be a disconnect between the inputs acquired under the 2KR program and the inputs needed by the recipient countries. It is clearly evident that the commodities procured under the 2KR scheme in Africa have been biased

106

towards pesticides (Tobin, 1996). In some countries, large machinery has been included on the list of items required when clearly there is no need for them (15 combine harvesters for Lesotho). In other countries, the items acquired on the list are not correlated with what the agricultural research institutes, in the particular country, recommend (ie: Guinea).

• *Inconsistencies in the use of the counterpart fund:* The current mechanisms of procurement and distribution, with lack of clear parameters on the use of the counter fund ('economic and social development'), has resulted in counterproductive uses.

These characteristics suggest that the aid program is not compatible with development efforts to achieve a market-oriented agro-input supply system.

What can be done to improve the effectiveness of the aid program ? One way to improve this aid would be to provide funds to the recipient government for direct procurement on commercial terms. This would be to establish a credit fund for providing loans to commercial firms to import fertilizers. Care would need to be taken to ensure the funds are used for their intended purpose. This approach would make the private sector responsible for the import transactions and thus have control over the technical specifications of the fertilizers procured, time of delivery, etc. Moreover, this approach will facilitate the development of a sustainable agro-inputs supply system in the recipient country by enabling emerging entrepreneurs to gain the knowledge and skills needed to handle fertilizer imports directly. If this or a similar change in approach to 2KR aid is not possible then it should be abolished.

In the meantime, the question then becomes how can recipient countries improve the effectiveness of the aid ? This can be done in a number of ways:
• *Create greater transparency and consistency in the preparation of the requirement list (request letter).* Governments needs to maintain consistency between the items to include on the request list (fertilizer, machinery and chemicals) and the agricultural development strategy of the country. All too often this has not been the case and items have been acquired that are not compatible with the agricultural strategy. ie:
  ⇒ The acquisition of a large quantity of machinery has, in some African countries, had a negative impact on employment creation with a substitution effect away from labor intensive technologies.
  ⇒ The acquisition of important quantities of chemicals has had negative environmental impacts and undermined the efforts to adopt an integrated pest management approach as an alternative to the improper and indiscriminate use of pesticides (Tobin, 1996, p.213).
  ⇒ Fertilizers acquired under the scheme need to be compatible with soil deficiencies and this should be considered in the development of the list.
• *Enhance the speed of completing relevant documentation (Request letters, Exchange of Notes, Banking Agreement and Authorization to pay).* To ensure that the inputs ordered under the scheme arrive in a timely manner, delays in completing the relevant documentation must be minimized to enhance the effectiveness of the aid. Receiving the fertilizer when it is needed (during the cropping season) reduces storage costs. Delays can also create pressure from interest groups who stand to gain from this program (fertilizer traders) to influence the choice of items on the list.
• *Ensure the sale of the 2KR aid (fertilizers) in the recipient country is at market prices.* Private traders in the recipient countries need incentives to participate in the market. Fertilizer sold at subsidized prices will inhibit private sector development. To minimize these distortions, fertilizer must be sold at the market price so that private traders can compete. This has been done successfully in Swaziland where 2KR fertilizers were well integrated into the domestic market. A

107

parastatal organization was selling 2KR fertilizers and South African fertilizers at the same depots, at the same time at competitive prices. One possible avenue for a market determined price is to auction 2KR fertilizer.

- *Eliminate subsidies on transportation and storage.* Subsidies on transportation and storage inhibit private sector development in these activities. If these subsidies continue, traders will have no incentives to store fertilizer or transport fertilizer to distant farmers.
- *Identify clearly defined uses of the counter part fund.* Like many aid programs, 2KR aid requires the development of a counterpart fund. While there has been some stipulation on how the counter part funds should be used (for 'economic and social development'), the use of this fund needs to be consistent with the overall agricultural development strategy and should be used in a transparent way. In some countries the use of this fund has been inconsistent with the progress of the agricultural strategy.
- *Uncertainty on government actions needs to be removed.* If the government is not consistent in its action with regard to 2KR aid in terms of selling inputs at market prices; reducing subsidies on transportation and storage; transparency and consistency in the preparation of the list then the uncertainty that is created could undermine the development of private sector activity in these input markets.

Coping with High Transportation Costs in Malawi

Prior to 1993, Malawi's fertilizer industry received extensive government intervention in pricing, importation and distribution. All fertilizers for small-holder farmers were procured and imported by the Small-holder Farmer Revolving Fund of Malawi (SFFRFM) and distributed by ADMARC which were both controlled by the government. The private sector's role was confined to serving the estate sector. Under this system, fertilizer use by small-holders tripled between 1980 and 1993, growing from 49 000 metric tons to 150 000 metric tons. This growth was, however, confined to the wealthiest third of small-holder households, typically those with access to agricultural credit, above average incomes and above average holding sizes (Conroy, 1996)[28]. Fertilizer consumption declined in 1993/94 following the collapse of the Small-holder credit system (SACA), the Supplementary Inputs Program then ensued, which distributed 32 000 tons of free fertilizer to drought affected households (Figure 6.13).

Removing fertilizer subsidies: Several measures were taken in 1993 following the Government announcements to phase out fertilizer subsidies and encourage private sector participation in fertilizer importation and distribution. The relationship between ADMARC and SFFRFM was terminated, with SFFRFM becoming an independent market participant. As a result of this more liberal stance, several private companies have emerged as importers/distributors. The market, once dominated by two parastatal organizations, now has at least eight major importer/distributions (dominated by Optichem, Norsk Hydro, Farmers World and Agora) and hundreds of private wholesalers/retailers (Chakravarti, 1997). In 1995/96 all fiscal and economic subsidies on fertilizer were completely removed leading to an increase in fertilizer prices varying between 200 and 300 percent. Consumption fell again to 71 000 tons, of which 23 000 was distributed for free by Government under the Supplementary Inputs Program (Chakravarti, 1997).

---

[28] The fertilizer sold in the late 1980s and early 1990s was largely financed by the Government as 60 to 70 percent of fertilizer was financed by credit which was not repaid.

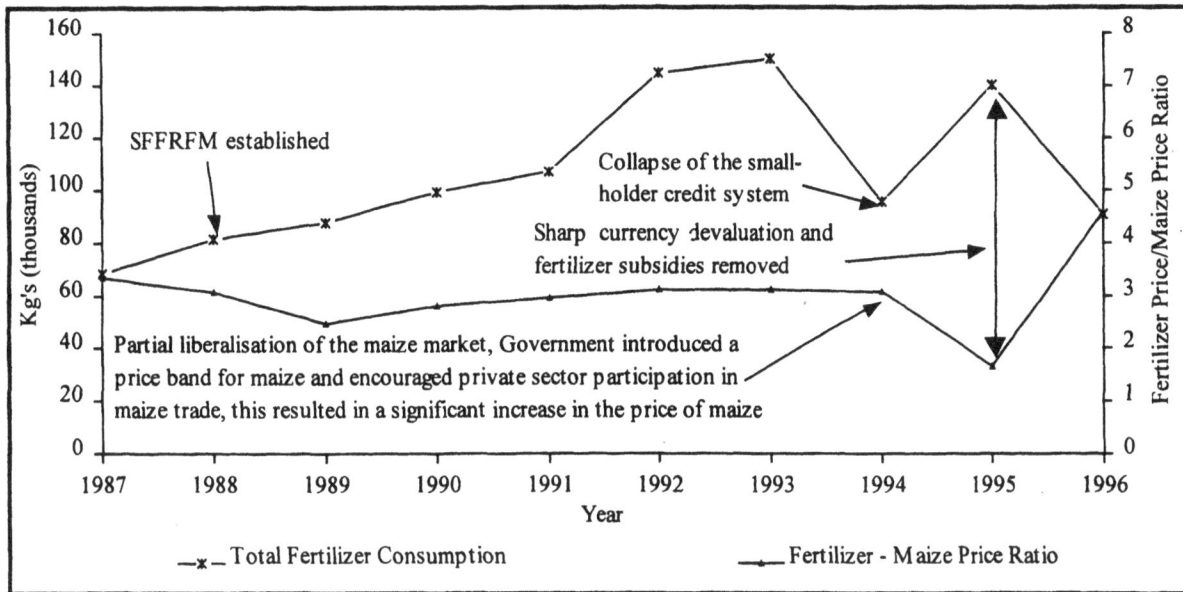

Figure 6.13: Fertilizer Consumption and Prices in Malawi.

The offsetting effects on the price increase of a more efficient low cost private sector have been slow to develop, although there are encouraging signs with efforts to minimize risks faced by traders. The distribution system of fertilizer has changed dramatically since 1994/95 when ADMARC's monopoly over fertilizer distribution was relinquished to an expanding private sector (Table 6.3). The major private companies have established a country wide network of branches and depots. As more companies enter the market, fertilizer mark-ups and profits should fall. The private trading companies continue to implement pan-territorial pricing, a system that was used by the government.

Table 6.3: Market Share of Fertilizer Distribution Agencies in Malawi.

Market participation	1994/95	1995/96	1996/97 (est.)
ADMARC	82	26	14
SFFRFM	0	18	14
Private Sector	18	42	63
Free Distribution	0	14	0
TAMA Scheme	0	0	9
Total	100	100	100

TAMA - the Tobacco Association of Malawi launched a fertilizer credit scheme.
Source: Chakravarti (1997).

Despite this change, fertilizer prices in Malawi have been viewed as being extraordinarily high (Conroy, 1997; Gisselquist, 1998). A recent study (Westlake, 1999) examining this phenomenon attributes these high costs to: the small size of the domestic fertilizer market; the country's landlocked position; inefficient transport systems in neighboring countries; the cost of distributing to a large number of rural outlets and the risks to importers and distributors, which stem from the uncertain level of effective demand. Within these factors, transportation costs play a dominant role. However, it appears that external land transport costs (port to Lilongwe) are extremely high and efforts to lower them could substantially reduce fertilizer prices in Malawi.

Table 6.4: Estimated Price Structure for Urea in Malawi Imported through Beira, March 1999.

Components of the Retail Price	% of Retail Price
FOB NW Europe ports, bulk	23.8
Ocean transport and insurance	7.0
Port charges, bagging an handling	9.2
Road transport Beira-Lilongwe	14.5
Inspection, bank, customs and clearing	1.8
Local transportation	0.9
Financing	7.0
Trading company share	25.1
Rural distributors margins	10.6
Retail price	**100**

Source: Westlake (1999, pg.9).

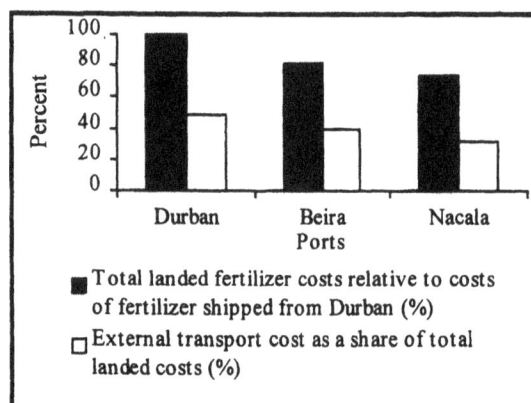

Figure 6.14: Fertilizer Transport Costs from Ports to Malawi.
Source: Conroy (1994).

Two ports in Mozambique, Nacala and Beira can be used by Malawi to import fertilizer. However, transport fees in Mozambique are particularly high and have been identified as one of the causes of the high costs shown in Table 6.4 (Westlake, 1999). Nacala is the closest port to Malawi but the lack of port capacity and outward freight results in exceptionally high shipping costs. Improving facilities at the Nacala port, namely, loading and storage facilities could indeed reduce costs. In earlier studies, Conroy (1994), using simulations of alternative costs from different ports, shows that external transportation costs can be significantly reduced if Mozambique ports are used (Figure 6.14).

Although Westlake (1999) suggests that internal transport costs are fairly low, these can be extremely variable and past studies have shown these costs to be high (Badiane, 1997, Conroy, 1994). These high internal transport costs may be reflected in the high share of the retail price taken by trading company's (Table 6.4). As other regional countries deregulate fertilizer trade, larger multi-country markets may develop, bringing larger bulk orders, more competition and lower prices to farmers. Indeed transportation is intrinsically a regional affair and market size in individual countries is usually too small to sustain modern infrastructure. Apart from developing regional transport services, another obvious means of reducing transport costs per nutrient is to use high analysis fertilizer.

Removing Price Controls and Trade Barriers in Zimbabwe

Zimbabwe's dualistic agricultural system, similar to Malawi's, has resulted in varying fertilizer application rates across farms. Fertilizer applications on large commercial farms are similar to those in developed countries. However, application rates on small-scale farms are relatively low and comparable with those in other African countries. Small-scale farmers account for 30 percent of total fertilizer consumption. Large regional differences in the use of fertilizer exist with variation across natural regions. Only 20 percent of small-holder farmers across the whole country apply fertilizer (Blackie, 1994). The application of fertilizers also varies significantly across crops within these two sectors. Smallholder farmers use 88 percent of the fertilizer on maize, while the commercial sector distributes it as follows: 33 percent on maize, 16 percent on tobacco and 14 percent on wheat.

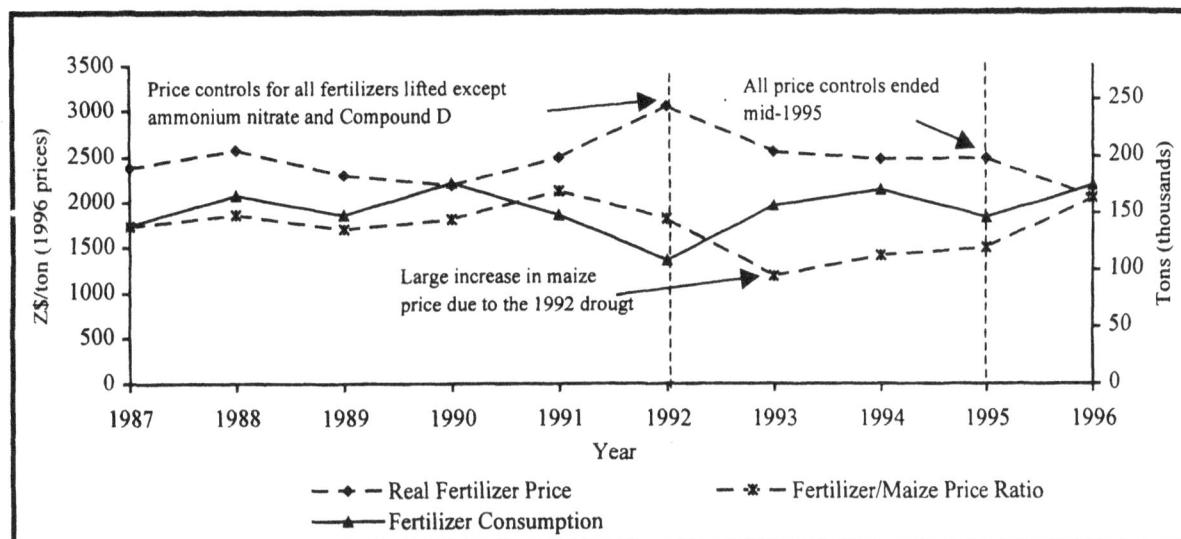

Figure 6.15: Real Fertilizer Price Trends in Zimbabwe, 1987-1996.
Source: Ministry of Agriculture, Zimbabwe.

As a result of policy measures adopted in the 1980s to expand rural infrastructure and introduce new small-holder credit schemes, fertilizer to smallholder farmers increased from 24 000 tons in 1974/75 to 90 000 tons in 1980/81 and reached a peak of 130 000 tons in 1986/87. This increase in fertilizer demand was heavily dependent on availability of credit and cash from crop sales, fertilizer prices, transport, timing of delivery and the distribution network (Takavarasha, 1995).

For several decades, the fertilizer industry has been dominated, in production and trade, by only a few companies (Sable, Zimphos, the Zimbabwe fertilizer company [ZFC] and Windmill) (Gisselquist and Rusike, 1998). As of 1990, the market opened up to increased competition (Table 6.5) but by 1996/97, the Ministry of Agriculture still maintained controls on fertilizer trade (stipulating which fertilizers can be traded) which continues to restrict competition. Domestic production provides more than half

Table 6.5: Market Share of Fertilizer Distribution Agencies in Zimbabwe.

Agency	1985	1990	1994
Co-ops	52	15	1
AFC large lending groups	21	13	9
AFC small lending groups	0	0	5
Traders	12	29	35
Manufacturers	15	43	50
Total	100	100	100

Source: Ministry of Agriculture Zimbabwe, 1997.

of all fertilizers used in the country. Fertilizer producers were protected by fertilizer import taxes with rates ranging from 0 to 5 percent and with a 10 percent surcharge on the finished products.

Fertilizer Policy Changes

In Zimbabwe real fertilizer prices reached their peak in 1992 after which all price controls, except on ammonium nitrate and compound D (subsequently removed in 1995), were lifted (Figure 6.15). Even though fertilizer prices declined from their 1992 value, urea sold in 1995 for US$300 against an international f.o.b. price of US$ 200 (Gisselquist, 1998). The higher fertilizer prices in

111

Zimbabwe reflect not only high transport costs for a landlocked country but protectionist taxes and relatively large margins in markets that are just starting to become competitive (Table 6.6).

The experience of the Zimbabwe fertilizer industry is represented in Table 6.6. The industry was highly regulated by the Ministry of Agriculture and trade was allowed according to nutrient composition. Arguments in favor of these trade controls is that restricting choices saves farmers from making wrong decisions, discounting their ability to manage. In 1992/3, the Government of Zimbabwe lifted price controls on all but two fertilizers, ammonium nitrate and compound D (recommended for maize), which were important to communal farmers; subsequently all price controls have ended. As of mid-1997, officials in the Ministry of Agriculture continue to manage fertilizer imports through permits, a non-tariff barrier.

Table 6.6: Summary of Impact of Reforms on the Zimbabwe Fertilizer Industry.

Pre-Reform Controls, Obstacles before 1990	Situation Early 1996
• access to foreign exchange, • import permits required, • mandatory government approval of fertilizer compositions, • price controlled, • parastatals offer subsidized fertilizer.	• removed, • no change, • no change, • removed, • removed.
Effects	**Effects**
• limited access to farmers technology, • few dealers, so that fertilizer not always conveniently available, particularly for small farmers, • limited choice of nutrients, farmers not able to match nutrients with soil deficiencies.	• Change in *industry structure*; formerly a protected public/private duopoly; with reform, ten new companies entered in 1995 and 1996, • *Technology transfer*; new entrants offer soil tests, made to order fertilizers, high analysis fertilizers, old companies respond with similar services, 260 new fertilizer products were registered between 1991-96. • *Prices*; competition is building; trading margin still high, • *Quantity;* no significant change; fluctuations in use dominated by austerity, drought and free fertilizer programs.

Source: Gisselquist (1998).

Several constraints facing the fertilizer industry have been highlighted by the Ministry of Agriculture (1997). These include: i) the industry is small, most of the machinery is old and generally not based on state-of-the-art technology; this means that it is not as efficient as technologies available in modern large-scale processing plants; ii) the products manufactured, particularly the compound fertilizers, are of relatively low analysis; which increases the cost of handling, transportation and storage per unit of nutrient and iii) for a long time the industry has been operating in a situation of oligopoly and is now in the process of adjusting to compete externally; marketing strategies of the fertilizer industry have been biased towards the commercial farming sector and have yet to develop facilities in the small-scale areas.

The Ministry of Agriculture's future strategies for the fertilizer industry consists of: i) encouraging more direct selling to farmers by manufacturers; ii) facilitating competition through tariffication, caution against export subsidies or any other unfair trading practice; iii) convincing small-holder farmers to set up their own fertilizer warehouses at strategic distribution points and iv) reviewing the legislation on fertilizer (Ministry of Agriculture, 1997, pg: 68).

Alleviating Inhibiting Regulations in Ghana

The use of fertilizer in Ghana remains low and has declined since 1990. This usage needs to increase if Ghana is to significantly improve agricultural growth. Past increases in agricultural production have been achieved mainly through area expansion which cannot be sustained in the long run. As with the other countries analyzed, most of the fertilizer in Ghana is consumed by cereal producers. Sixty-four percent is used for maize, 10 percent for rice, 11 percent for vegetables and less than 1 percent for cocoa and coffee.

The procurement and distribution of inputs in Ghana was liberalized in 1992. Subsidies, which used to be about 65 percent in 1985, were gradually reduced to zero (Figure 6.16). In this liberalized market, fertilizer prices have increased at a much faster rate than output prices at the farm level. Comparison of import prices (cost and freight) of related fertilizers in Côte d'Ivoire in 1993 was US$ 200/mt while in Ghana it was US$260/mt (Gisslequist, 1996). The removal of fertilizer subsidies and the depreciation of the Cedi resulted in significant fertilizer price increases after 1993. Fertilizer consumption in Ghana peaked in 1981, declined in 1984 and has since remained relatively constant (The decline may have been provoked by the severe drought in 1983). The reduction in fertilizer use can also be attributed to the problems with supply of fertilizers to farmers, and until this problem is dealt with, fertilizer consumption is not likely to recover (Donovan, 1996).

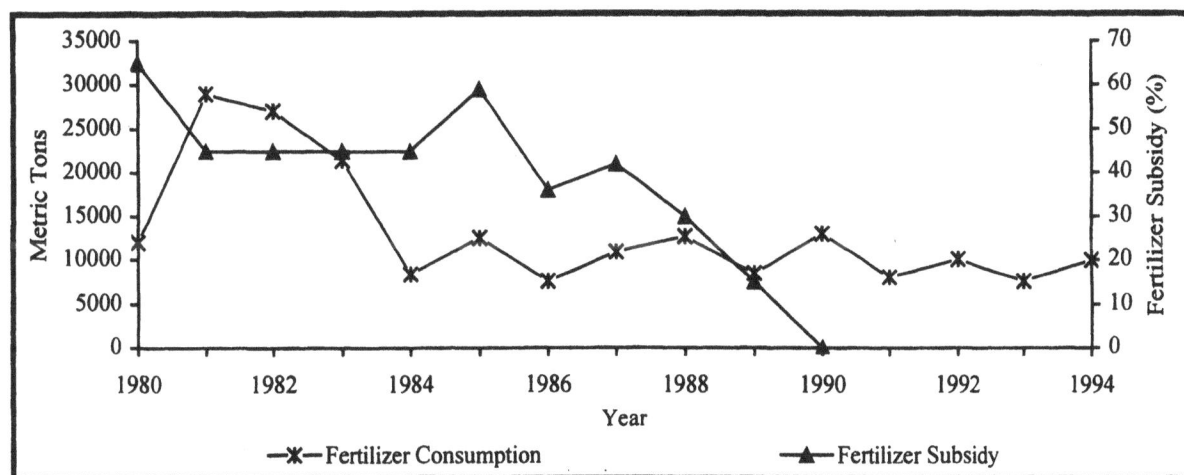

Figure 6.16: Fertilizer Consumption and Subsidies in Ghana.

The privatization of the supply and distribution of fertilizer in Ghana was meant to improve the fertilizer availability to farmers. However, there are still complaints about fertilizer unavailability in certain localities. A small number of active dealers complain of low margins (Bumb, 1994). Four firms entered the fertilizer import and distribution business in 1992. By 1994, one importer with financing from manufacturers had acquired a major share of the market and was reaping off the benefits of the economies of scale in importing. There may be some evidence that the development of this oligopsony has had negative effects on prices. It is generally perceived that privatization has not resulted in increasing competition, providing better farmer services, offering lower relative prices and expanding the use of fertilizers by farmers. One possible reason for the development of the oligopsony is the economies of scale in importing.

113

Regulatory Issues

Continuing trade controls restrict fertilizer market entry and competition, contributing to high margins and prices. The restrictive list of fertilizers that can be imported into Ghana denies Ghanaian farmers the opportunity to use nutrient-wise, more cost-effective fertilizers like Diammonium Phosphate. High prices have also been attributed to some of Ghana's fertilizer regulations, which cut Ghana's farmers off from fertilizer markets in neighboring countries, forcing importers to consider Ghana as a separate and small market (Gisselquist, 1998). Private traders are not able to bring in fertilizers without prior government approval, this involves time consuming and costly procedures to satisfy multiple agencies. To improve fertilizer trade, reduce prices to farmers and increase the range of fertilizer technology available, Gisselquist (1998) highlights several considerations for trade liberalization.

- Allow importation of all kinds of fertilizers, stipulating only that products imported as fertilizers should not be harmful if applied as fertilizers. There is currently a list of fertilizers that importers are allowed to bring in and sell.

- Do away with requirements for prior approval or clearances from government agencies before importers can order or receive imported fertilizers. Traders require prior approval for fertilizer imports (before importing and before clearing goods from the port or border).

- Relax packaging regulations and specifically allow fertilizers that are packaged according to regulations in all neighboring countries.

- Enforce truth-in-labeling (to control fraud) and prohibit the sale of anything as fertilizer that could be harmful to the soil. This involves making rules for sampling and setting ranges of allowed variance around stated compositions for various nutrients. At the same time, regulations can generally disallow any component that might be harmful.

Similar recommendations for fertilizer procurement and distribution in Ghana were suggested by IFDC[29]. They include: i) collaborating with Côte d'Ivoire in combined importation; ii) relaxing current import restrictions on the specification of fertilizers in order to allow importation of nutrient wise more cost effective blends of fertilizers such as diammonium phosphate; iii) stabilizing the Cedi; iv) assisting importers in accessing adequate levels of credit for importing fertilizer; and v) evaluating rock phosphate supply in the Keit Basin in Ghana and in Togo (Gerner, 1995).

Learning from the Ethiopian National Fertilizer Project

Until 1992, the fertilizer market in Ethiopia was totally controlled by the state owned parastatal named the Agricultural Input Supply Corporation (AISO), subsequently renamed the Agricultural Input Supply Enterprise (AISE). In 1992, the New Marketing System (NMS) was set up to liberalize the fertilizer market and create a multi-channel distribution system. The monopoly of the state-owned parastatal was removed and the private sector was then able to engage in the importation and distribution of fertilizer. Since then, two firms have joined the market of fertilizer import and distribution. Fertilizer subsidies and pan-territorial prices have been removed and, with the currency devaluation, fertilizer prices have increased by 250 percent since 1991. In 1993, only 16 percent of farmers in Ethiopia used fertilizer; this number increased subsequently to 25 percent (Sodhi, 1999).

---

[29] The IFDC report suggests that the costs of fertilizer in Ghana could be easily reduced if "..it actively promotes the importation and use of straight fertilizers, as well as by allowing the importation of blended fertilizers from Côte d'Ivoire."

Although access to fertilizer is thought to have improved as a result of the input market liberalization, Demeke *et al* (1998) identified four major problems affecting demand from their analysis in 1998:

a) Retail markets are poorly developed, thus limiting farmer's access to fertilizers. Small-scale wholesalers' and retailers' participation has been weak which can largely be attributed to the way credit is allocated (access the credit), the removal of fertilizer subsidies and the unattractive wholesale price fixed by the government.

b) The system of credit disbursement to farmers discourages competition and leads to market concentration and uncertainty for potential new entrants in fertilizer distribution. The responsibility of credit disbursement and collection has been transferred from the banks to the regional governments. The regional governments estimate their fertilizer credit requirements and sign loan agreements with the bank. This process creates severe market distortions as only firms favored by the authorities are nominated as suppliers. For many years excess supply has prevented further private sector entry. In 1996 only 60 percent of the total amount of fertilizer available to farmers was actually sold. Large carry over stocks created additional costs due to storage fees and interest charges.

c) Fertilizer dealers/agents were not allowed to buy fertilizer from input suppliers of their choice, they operate as commission agents of importers and are unable to establish themselves as independent operators. Introduction of licensing of fertilizer dealers would alleviate this problem.

d) Regulation of prices. Although retail prices have been deregulated, the wholesale price is fixed by government.

Future reforms should address these issues. Despite these constraints fertilizer sales have almost tripled since 1991, increasing from 110 000 tons to 300 000 tons in 1998. Credit recovery from farmers is about 90 percent and cash purchases have increased from zero to thirty percent. Food-grain production has almost doubled, increasing from 6.5 million tons in 1991 to 12 million tons in 1998. Several factors caused this improvement.

In 1993, the NGO, Sasakawa Global 2000 and the Ministry of Agriculture began a joint program to demonstrate that substantial productivity increases could be achieved when farmers were given appropriate extension messages and agricultural inputs were delivered on time at reasonable prices. By 1998, five years after the beginning of the program, the number of demonstration plots had increased to 2.5 million. Farmers used their own land (half-hectare plot) and were provided credit, inputs and extension. There has also been increased Government, NGO, Donor and regional governments collaboration through annual three-day National Fertilizer Workshops, which provide a platform to discuss current issues (supply and demand, etc.) in the fertilizer industry. There is clear evidence that this increased use of fertilizer and seeds has raised yields significantly (Howard *et al*, 1999) and improved the profitability of maize and teff production.

Lessons Learned from the Relative Success of the Ethiopian National Fertilizer Project

Several lessons have been elaborated by Sodhi (1999):

- Economic reforms are not sufficient; the transition to a free input marketing system needs careful management and monitoring.
- Government commitment to reform is critical.
- The private sector needs time to build up knowledge and confidence in reform programs.

- Collaboration among government, private sector, NGOs and donors is critical to success.
- Strong research-extension support is needed to promote efficient and environmentally safe fertilizer use.
- Quality control of fertilizer ingredients is critical to any successful fertilizer program.

SUMMARY AND CONCLUSIONS

Declining soil fertility has been highlighted as a major reason for slow growth in food production in Sub-Saharan Africa, a trend which could be reverse by increasing fertilizer use. Fertilizer applications on the continent are very low (on average at one fifth of those used in other [non-irrigation] developing countries) and application rates have remained relatively constant since 1980. Thus, there remained great opportunities for expanding agricultural production through improved fertilizer use.

- Fertilizer adoption and use intensity are determined by the profitability of it's use, human resource availability, financial liquidity, household assets, market access and extension services. This list of explanatory variables highlights the complementarity between prices and non-prices factors.
- Prices remain an important component of the determinants of profitability and explanations for the cross-country differences in the ratio of fertilizer prices to world prices include the distance from international fertilizer markets, the quality and quantity of roads in the country, exchange rate policies, fertilizer import tariffs, and the volume of fertilizer used. Internal and external transportation problems exist even in coastal countries. The low volumes purchased reduce economies of scale in importing since 1990.
- Real domestic fertilizer's prices increased in all the African countries analyzed in this study which can be explained by the increase in world prices, devaluation of the domestic currency and the removal of fertilizer subsidies. In several countries, real fertilizer prices have increased by over 100 percent.
- Once the effects of currency devaluations and world price increases are taken into account, six of the eleven countries analyzed have reduced the real price of urea to farmers (Zimbabwe, Côte d'Ivoire, Kenya, Burkina Faso, Togo and South Africa). This suggests that there has been an improvement in the 'efficiency' of these fertilizer markets. In five countries (Malawi, The Gambia, Ghana, Nigeria and Cameroon), the real fertilizer prices have increased.
- Tariff and non-tariff barriers continue to inhibit private sector activity. Limitations on the type of fertilizers that can be imported, rigid (clearance) approval requirements and packaging regulations restrict private traders. Experience proves that removing these restrictions encourages private trade and ultimately reduce fertilizer prices. This measure must be supplemented with the introduction/enforcement of a truth-in-labeling.
- Many African countries continue to rely on fertilizer aid which makes them vulnerable to supply uncertainties, limitations on product choice and disruptions of domestic fertilizer markets.

In order to encourage private sector activity, economic reforms need to be complemented by carefully managed and monitored fertilizer systems. Moreover, the credibility of fertilizer reforms needs to be maintained, and collaboration between government, private sector, NGO's and donors encouraged. Quality controls should also be strictly imposed. Government investment in roads (both quality and quantity), prudent exchange rate policies, the removal of inhibiting trade restrictions and less dependence on fertilizer aid will help to encourage private participation in these input markets.

116

PART IV

REMOVING BARRIERS TO IMPROVE AGRICULTURAL INCENTIVES IN SUB-SAHARAN AFRICA

7. POLICY CONSIDERATIONS

INCENTIVES AND THE POLICY LANDSCAPE IN AFRICAN AGRICULTURE

Over the past several decades, the agricultural policy landscape in Sub-Saharan Africa has changed significantly. From a once highly interventionist environment, the policy climate has been transformed into one of fairly wide-spread market liberalization, although the contrasts of change among countries remains extreme. The improved policies have been accompanied by high agricultural growth rates for many African countries. Between 1990 and 1997, 25 countries experienced real agricultural GDP growth rates of over 2 percent while 12 of these countries experienced growth rates of greater then 4 percent. A large portion of this growth can be explained by the changes in the incentives faced by farmers. Over this period, world commodity prices were more favorable, macroeconomic policies continued to improve and the producer's share of the border price increased in many countries. The improved policies have improved technical and allocative efficiency and induced greater agricultural growth.

The macroeconomic policy environment: Many African countries have improved both their macroeconomic and agricultural policies over the last two decades. The large exchange rate overvaluations of the 1970s and 1980s have been reduced and inflation and budget deficits have been lowered in many countries. There are remaining curbs on imports (tariff and non-tariff barriers) but there appears to be renewed energy to promote exports. Achievement of the macroeconomic targets remains mixed across countries with a fairly high degree of macroeconomic instability experienced by many countries (Zimbabwe, Ghana and Nigeria) in the mid-1990s. Some of the key elements of the macroeconomic environment can be summarized as:

➢ Widespread reduction in overvalued exchange rates with a movement towards equilibrium levels. Current parallel market exchange rate premiums in most countries are low. Francophone Africa maintains a fixed exchange rate, pegged against the French franc. The CFA franc devalued by 50 percent in 1994.

➢ Fiscal policies have improved with lower fiscal deficits in most countries. Fifteen countries (out of 26 analyzed in the study) have fiscal deficits of less than 4 percent of GDP.

➢ Monetary policy has shown mixed results with limited improvement since 1990. Nine countries have inflation rates of 20 percent or greater and seigniorage remains high in many countries.

➢ Financial regulation remains weak in many countries (e.g. Zimbabwe and Nigeria)[30] and this is a key area for improvement.

➢ The credibility of economic reforms in Sub-Saharan Africa is low relative to other regions of the World (World Bank, 1997) and Brunetti *et al* (1998) have shown that this lack of credibility has had a negative impact on investor confidence with corresponding negative externalities on growth.

---

[30] Lessons from the Asian crisis suggest that the weak financial sector was the main contributor to macro-instability (IMF, 1998). In some cases extensive freeing up of financial markets could contribute to increased macroeconomic instability.

The export crop sector: Favorable world prices and the devaluation of domestic currencies have contributed to an increase in real domestic export crop prices across almost all African countries. In several cases, an improvement in the efficiency (price transmission) of the market has resulted in further price gains (e.g. Uganda, Nigeria, Tanzania, Togo, Madagascar, Zimbabwe and Malawi). Exchange rate deprecations have been fully passed-through to farmers in markets where countries have adopted open trade policies (e.g. vanilla in Madagascar, cotton in Zimbabwe, tobacco in Tanzania and Malawi and cocoa in Nigeria). In markets where there has been continued state interventions (pricing, distribution and marketing) this pass-through has been inhibited. Nine traditional crops continue to dominate exports from the continent and Sub-Saharan Africa's world export share for five of these crops (bananas, cotton, sugar, tobacco and tea) has increased in the 1990s. Non-traditional export crop production has become more widespread with an increase in the production of cut flowers, citrus and deciduous fruit, pulses, live animals and cashew.

The recent evolution of policies, prices and markets, particularly since the 1980s, has resulted in a general improvement in the export crop production incentives faced by many farmers in Sub-Saharan Africa.

➢ There has been widespread adoption of market liberalization policies in many countries (eg: Uganda, Tanzania, Mozambique, Nigeria and Zimbabwe), particularly in Eastern and Southern Africa.

➢ With the removal of marketing boards, the private sector has actively taken over the marketing responsibilities in several countries (eg: coffee in Uganda and cotton in Zimbabwe).

➢ Marketing boards (eg: cocoa in Ghana), *Caisse de Stabilization* systems (eg: cocoa and coffee in Côte d'Ivoire) and parastatals (cotton in West Africa) exist in many countries. Marketing boards and parstatals typically set pan-territorial and pan-seasonal prices and control the physical handling of the crop. Under the *Caisse* system prices are administratively determined with purchasing and selling prices set at each stage of internal commercialization in an attempt to stabilize prices.

➢ Although the export crop sector was heavily taxed, there was limited public investment back into rural areas (this is different from other developing regions such as East Asia where agriculture was heavily taxed, but this was combined with significant public investments). The explicit taxation from overvalued exchange rates in Africa has been significantly reduced, and in many countries has been eliminated.

➢ Access to foreign markets varies greatly between countries in Africa. Although OECD import tariffs on African products are generally low, there are significant non-tariff barriers such as strict sanitary and phytosanitary requirements. Many countries continue to receive preferential trade treatment under the Lomé Convention and the Generalized System of Preferences, receiving export prices higher than world market prices (Botswana beef, Mauritius sugar). High external transport costs have often eroded the benefits of these higher prices. For some crops, however, special taxes are applied to imports in several European markets (ie: coffee).

➢ Real world commodity prices are characterized by a long-term downward trend. Fluctuations around this declining trend remain substantial (booms and busts).

120

The food crop sector: Liberalization of food markets has been widespread in Sub-Saharan Africa. Marketing boards have been dismantled and prices have been liberalized in most countries. The impacts of these reforms have been greatest for consumers as food costs have been lowered. A general overview of the current policy landscape for the food crop sector in Africa can be broadly summarized as:

➢ Over the last several decades, Governments of West African countries have intervened less in food markets than their Southern African counterparts.

➢ Despite liberalization efforts in Eastern and Southern Africa, many countries continue to intervene in grain markets. Back-sliding of reforms has been a common trend in many countries (Malawi, Zimbabwe and Tanzania). Food security issues are high on Government agendas in African countries, which together will recurrent droughts, has increased pressure for state intervention in these markets.

➢ Several countries continue to set prices such as administer floor prices (Malawi) and ceiling prices on foods (maize meal in Zimbabwe).

➢ Strategic grain (both as grain and cash) reserves have been developed in many countries.

➢ In several African countries, licenses for grain trade are required which have impeded private imports and exports, reducing their potential to stabilize food supplies and prices.

Other key issues in the sector include:

➢ Women account for a much larger proportion of the time spent on food production than men. Given the multitude of tasks women undertake in the household (collecting firewood and water, cooking, etc.) and the significant labor constraint they face, policies to ensure food security will require the development of appropriate labor saving technologies which cover the full range of their domestic and economic tasks.

➢ Private sector response to reforms has been mixed. There has been large market entry into marketing niches with low entry costs. However, mobility barriers continue to restrict private sector entry into all niches along the marketing chain.

➢ High transportation costs have been a source of high food prices in many African countries. Private investment in transportation and storage has been inhibited by the continuation of policies such as pan-territorial pricing structures.

Fertilizers: Fertilizer application rates in Sub-Saharan are the lowest in the world and have stagnated since 1975. Increasing their use is critical to reverse the decline in soil fertility. This declining trend has been highlighted as a major reason for slow growth in food production in Sub-Saharan Africa (Sánchez *et al*, 1995). Government interventions in the fertilizer markets have been reduced significantly over the last decade.

➢ Fertilizer subsidies have been removed in nearly all Sub-Saharan African countries.

➢ This, together with currency devaluations, have caused real fertilizer prices to rise significantly, in some cases by 100 percent. With these high costs, several countries face internal pressure to backslide on previous reforms with a reintroduction of fertilizer subsidies.

➢ Many African countries rely almost exclusively on fertilizer aid to meet their domestic requirements. This aid dependence has caused uncertainties in supply, product choice is limited and domestic fertilizer markets are disrupted.

➢ The private sector response to fertilizer market liberalization has been weak. Generally a few large private firms dominate the market in SSA countries.

➢ One contributing factor to this limited emergence is that trade restrictions are still widespread with high tariff and non-tariff barriers. In some countries these non-tariff barriers take the form of restrictions on the types of fertilizer that can be imported, stringent clearance requirements for imports and specifications on who can import.

➢ Fertilizers volumes traded remain extremely small which has reduce bargaining power and raised unit transport costs.

Ranking of the Agricultural Policy Stance for African Countries, 1996-97: The rankings provided below show countries which are deemed to have better price policies first. While rankings by definition always require a first and a last, it would be more sensible to refer to an upper, middle and a lower group. This approach should be used in the interpretation of the results.

Price and non-price factors: Both price and non-price factors affect these policy scores. For example, the producer's share of the border price, as seen in chapter 5, is not only affected by agricultural policy, but also by variables such as the development of rural infrastructure which is determined by infrastructure policy. Several non-price factors were also used to develop the fertilizer policy scores. These include non-tariff barriers such as the extent of government controls and regulations through import permits and restrictions on types of fertilizer to be imported, etc. (see Table A9). All of these factors have an effect on farmers and private sector incentives to produce and invest.

Macroeconomic Policy Scores

Cameroon
Senegal
Benin
Mali
Uganda
Burkina Faso
Côte d'Ivoire
Tanzania
Kenya
South Africa
Togo
Malawi
The Gambia
Nigeria
Ghana
Zimbabwe

0 50 100

Producer's Share of the Border Price for Export Crops

Nigeria
Kenya
South Africa
Malawi
Uganda
Zimbabwe
Cameroon
The Gambia
Tanzania
Côte d'Ivoire
Senegal
Mali
Togo
Ghana
Benin
Burkina Faso

0 50 100

Food Crop Policy Scores

Ghana
Senegal
South Africa
Cameroon
Benin
Togo
Uganda
The Gambia
Mali
Tanzania
Malawi
Côte d'Ivoire
Kenya
Zimbabwe
Burkina Faso
Nigeria

0 50 100

Fertilizer Policy Scores

South Africa
Uganda
Zimbabwe
Malawi
Côte d'Ivoire
Kenya
The Gambia
Mali
Tanzania
Cameroon
Ghana
Burkina Faso
Nigeria
Senegal
Togo
Benin

0 50 100

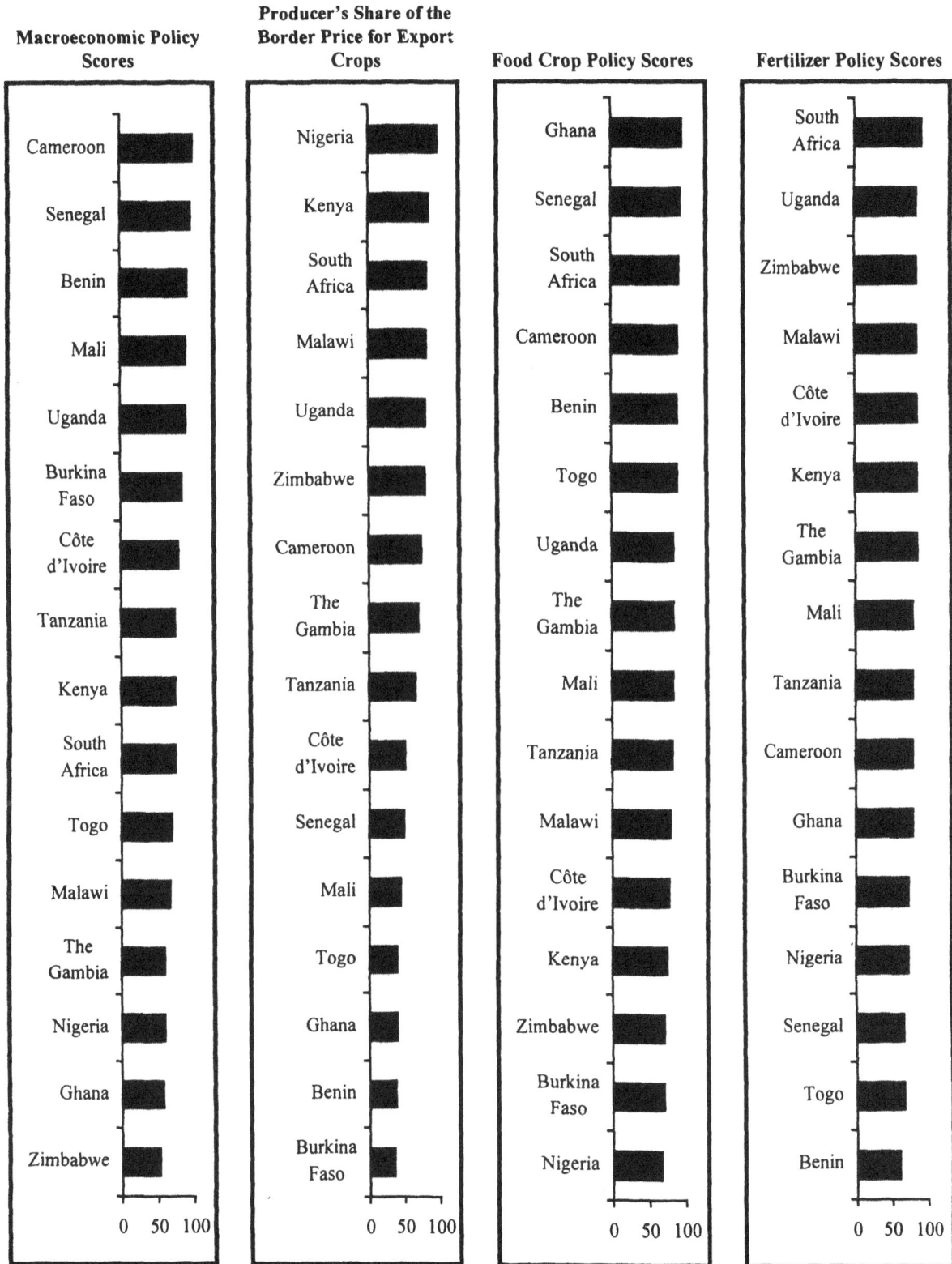

Figure 7.1: Ranking of Price Policy Indicators for African Countries, 1996-97
For a brief description of the state of policies in each of these African countries, see the county profiles in chapter 8.

REMOVING MARKET AND PRICE BARRIERS TO IMPROVE AGRICULTURAL INCENTIVES

The study has shown that the external environment and policy regimes have improved in a many African countries. However, agricultural incentives in a significant number of countries continue to be inhibited. Transportation costs remain high, traded volumes remain low, private agents have not entered sufficiently into input and output markets and the degree of market development and competition remains low. Tariff and non-tariff barriers to agricultural trade continue to be high inside Africa and in global markets, with improving access to OECD markets being a major challenge. Thus, evidence suggests that there continues to be significant room to improve agricultural incentives. This improvement will require focused attention on the macroeconomic environment, trade and market access and on domestic agricultural and rural policies.

THE MACROECONOMIC ENVIRONMENT

Stabilizing Macroeconomic Policies

Macroeconomic instability remains high in several African countries which has had an adverse effect on investor confidence and growth. To maintain stability, countries need to maintain realistic exchanges rate, low inflation and low budgets deficits. While these are first generation requirements, they need to be supplemented with prudent financial market rules and a high level of reform credibility.

Enhancing the Institutional Framework and Credibility of Rules

Based on the credibility index from the 1997 World Development Report, which reflects an overall indicator of the reliability of the institutional framework as perceived by entrepreneurs (WDR, 1997, pg. 5), Sub-Saharan Africa has the second lowest ranking (after the Commonwealth of Independent States) when compared to other regions of the world. Encouraging private sector entry into liberalized markets (and maintaining their participation) will require a significant improvement in the credibility of reforms. About 60 percent of entrepreneurs in Sub-Saharan Africa reported that unpredictable changes in rules and policies seriously affected their business. This is critically important, as this has surely inhibited private sector development in African economies. As stressed by the report, 'countries need markets to grow, but they need capable institutions to grow markets' (WDR 1997, p. 38). Institutional arrangements that foster responsiveness, accountability and the rule of law need to be developed.

TRADE AND MARKET ACCESS

Removing Domestic Trade Barriers

Trade barriers are still high in African countries. African import tariffs average about 25 percent which is three time higher than those of the fast growing exporters and more than four times the non-OECD average (Ng and Yeats, 1996). In agriculture, while tariffs on manufactured fertilizers have been reduced and are relatively low, large tariffs on transport equipment, machinery and agricultural materials and high taxes on petroleum products and lubricants still exist. These tariffs are an additional direct cost to exporters who use these imports for intermediate inputs in production and act as a cost disadvantage, eroding competitiveness and inhibiting growth. Tariff barriers need to be significantly reduced to enhance growth and reduce the price of inputs to farmers. In the case where countries have both fertilizer subsidies and trade barriers, appropriate reform sequencing may be a useful tool to off-set the large fertilizer price increases that result from subsidy removal, ie: removing these import barriers and subsidies at the same time could leave domestic fertilizer prices essentially the same. While these trade barriers are high, commitments by African countries to reduce agricultural import tariffs remains low. Under the Uruguay Round Commitments, most African country tariffs are bound between 50-100 percent (Ingco and Townsend, 1998).

Non-tariff barriers are also widespread across Africa, restricting market entry and competition. Several countries have a restrictive list of fertilizers that can be imported which, in some cases, has inhibited market determination of fertilizer composition (Gisselquist, 1998). Private traders in many countries are not able to import fertilizers without prior government approval involving time consuming and costly procedures to satisfy multiple agencies. These non-tariff barriers need to be removed to allow increased private sector activity in these markets.

Improving Access to Foreign Markets

Agricultural and agro-industrial growth cannot occur if producers are confined to local markets, thus improving Africa's access to foreign markets will be essential to stimulate growth. There are concerns that this access, particularly to OECD markets has been inhibited (World Bank, 1997). Three principle concerns raised include high external transportation costs, import barriers and strict sanitary and phytosanitary requirements.

External transport barriers could be reduced with regional cooperation on maritime transport, ensuring that the African shipper councils adhere to the goal of promoting foreign trade and ensure active monitoring of costs along the transport chain to identify and take action where transport costs are excessive (World Bank, 1997a). OECD import tariffs have not caused severed market access restrictions for Africa. In fact, many African countries continue to receive Preferential Trade Agreements. Non-tariff barriers, such as strict sanitary and phytosanitary requirements, remain a potential barrier to expand trade into foreign markets. African countries need to build up capacities in the public and private sector to apply these increasingly rigorous standards. These will include building effective systems to control or eradicate plant and animal diseases and to ensure the safety of exports.

As world agricultural trade becomes increasingly influenced by international trade agreements under the WTO, Africa will need to develop stronger bargaining power in these international negotiations. Historically, African countries have played a small role in this arena. During the recent Uruguay Round negotiations on agricultural trade, the major players continued to consist of the developed countries, however, a number of African countries, such as Nigeria, Senegal and Tanzania managed to actively participate in the Round. The motivation of these countries during the negotiations was mainly to safeguard old preferences or obtain compensation for potential adverse effects from higher world food prices on their import bills. As of May 1995, 38 countries in Sub-Saharan Africa (including 24 least-developed countries) had signed the World Trade Organization (WTO) Agreements and had submitted their schedules of Uruguay Round commitments in agriculture, industry and services. The WTO trade negotiations in November 1999 provide an ideal platform for African countries to voice their concerns about market access restrictions. An effective approach may be to empower larger customs unions or free trade zones to put forward Africa's case (SADC and UEMOA) (Ingco *et al*, 1999).

Fostering Regional Integration

Consistency in the supply and demand of agricultural products and inputs is a requirement for the development of these markets. The private sector will only enter markets where they are confident that products will be available to trade. Due to the small trade volumes of crops and agricultural inputs (particularly fertilizer) in African countries, economies of scale are not realized (annual fertilizer requirements for many African countries are not large enough to fill a ship, thus increasing unit transport costs). These higher transaction costs will lower producer prices and raise inputs (fertilizers) prices for farmers.

Regional integration may provide an avenue to ensure both consistency and larger volumes of supply and demand in product and factor markets, especially for fertilizer. The pooling effect of the rather small markets should enabling more bulk production and purchase of raw materials. Similarly, larger volumes allow stronger bargaining powers on world markets. However, the pooling of these markets is difficult due to the fragmented policy regimes among African countries. For countries to benefit from region trade and pooled markets the rules of the game (norms and procedures etc) must be standardized. This will indeed be important for fertilizers and for certified seeds. These larger input and output markets would be able to attract major multinational agro-business firms to invest in both production and processing.

Coping with International Commodity Price Decline and Fluctuations

Most African countries continue to depend on a few agricultural export crops. Over the last century, real world agricultural commodity prices have shown a declining trend, with some concerns that price volatility has increased. African development strategies need to recognize this 'stylized fact' and include coping mechanisms in their development plan.

Diversification is often suggested to deal with declining prices and is a strategy more suited to cope with the secular price trends as opposed to cyclical trends. Diversifying the export base away from the traditional export crops requires significant investments to gather information on new markets and products. Although real world prices for non-traditional export crops have declined, their rate of decline has been slower than the corresponding decline for traditional exports. Nevertheless, the downward trend in world commodity prices is likely to continue in the future, which suggests that Africa must produce agricultural commodities at lower cost to prevent further erosion of its share on world markets. This is only possible via continuous technological innovation and adoption. In this regard, biotechnology will inevitably become more important globally. If African countries are to maximize the benefits of this new technology they will need to build their own national capacities to identify opportunities, gain access to appropriate technologies and evaluate the risks associated with their use (Byerlee and Gregory, 1999).

In Africa, the common coping mechanisms have been variable export taxes and internal and external price stabilization mechanisms (*Caisse d'Stabilization* and Marketing Boards) with many countries being members of the International Commodity Agreements. Experience suggests that export taxes have been set too high and that revenues generated have not been effectively utilized. Price stabilization policies, under the *caisse* systems, have been difficult to maintain due to economic and political pressures and international commodity agreements have proved to be unsuccessful over the longer term.

When suggesting coping strategies cyclical trends need to be differentiated from secular trends. Common prescriptions to cope with income and foreign exchange changes due to cyclical price booms include (Varangis *et al* 1995) : i) *Don't overspend or over-commit.* Some of the revenues should be used to build up international reserves or to reduce debt (Botswana has done this fairly successfully with its diamond revenues). ii) *Adopt prudent monetary and fiscal policies.* A typical problem is that fiscal spending plans get based on high revenues (boom years), and when these fall, the plans are maintained through distortionary borrowing. External debt could also be reduced with the extra revenues generated. iii) *Hedge.* Market-based hedging instruments should reduce price uncertainty (McIntire and Varangis, 1998). A commodity exchange has been developed in Zimbabwe and South Africa with a fairly effective use of forward contracting; iv) *Avoid high export taxes.* Stabilization taxes are often imposed in commodity boom years (ie: coffee in Uganda) and v) *Avoid cartels.* International commodity agreements have not been successful at raising prices and they have fail to constrain supply effectively.

In recognition of the importance of these cyclical trends to developing countries an *International Task Force (ITF) for Commodity Risk Management in Developing Countries* (led by the World Bank), has been set up to explore sustainable and effective ways to assist developing countries to better manage their vulnerability to risks associated with commodity price fluctuations. The task force will investigate critical issues, and propose new market-based approaches to manage commodity risks.

DOMESTIC AGRICULTURAL PRICE AND NON-PRICE FACTORS

Removing Remnants of Marketing Boards

While policy reforms in many African countries have resulted in increases in the producer's share of the border price, interventions in several export crop markets continue to constrain producer prices. *Caisse de Stabilization* and marketing boards intervene in physical handling, price setting, taxation and marketing costs and margins. Producers in these markets are typified by a low producer's share of the f.o.b. price. There are some concerns that removal of these structures will result in a loss in bargaining power on export markets and quality standards will be reduced, all leading to lower prices on world markets. As yet, there does not seem to be any clear evidence that corroborates this hypothesis. Apart from cocoa, market shares are not significant enough to allow marketing boards to exercise any market power on world markets. The domestic price increases from the removal of marketing boards appears to mitigate any world price declines resulting from the loss in bargaining power, should there be any. Quality standards, however, need to be maintained in these more open markets. There is a need to encourage producers to establish professional organizations which could help in increasing their bargaining power and quality standards, governments can play a role in recognizing and legitimizing the quality norms.

Removing Excessive Agricultural Taxation and Enhancing Public Rural Investment

As extreme taxation (via, for example, marketing boards) is dismantled, the question arises as to how the sector should be taxed in the future. As the major sector in many African economies, agriculture will have to contribute to government revenues. The key principles that must be applied to future agricultural taxation are nondiscrimination, minimization of negative efficiency impacts, effectiveness of fiscal capture and capacity to implement (Binswanger *et al*, 1999).

Agricultural taxation, where possible should be integrated into general value added, profits, income and wealth taxation. Efficiency losses must be minimized via the minimization of output and input taxes. Consumption taxes (such as sales and value added taxes) have the advantage that they don't effect the efficiency of production. Although consumption taxes have become a more popular form of taxation in Africa, administrative capacity to implement seems to remain limited in many Sub-Saharan African countries. The capacity to implement these tax systems needs to be strengthened and, in many cases, will have to be built up over many years during which little revenue will be generated. In this context, export taxes may be justified until the administrative capacity is developed. It appears that where export taxes are justified as substitutes for income tax in Africa, rates should be reduced substantially. The same its true for input taxation.

Despite these high levels of past taxation, public investment (public services and infrastructure) in rural areas was poor. If indeed these high taxation levels were complemented with significant public investment (as in Asia) then agricultural growth would not have faired so poorly in Africa. Fiscal decentralization will improve the level and efficiency of these public investments but the progress of decentralization and devolution of resources and responsibility in Africa remains slow.

Improving Transportation Infrastructure

One area of high priority for public investment in rural areas is developing rural infrastructure. High transportation costs in Africa are one of the key contributors to low producer prices for export crops and inflated import prices for agricultural inputs. These costs are exacerbated for landlocked countries which are separated from ports by long distances. Road density and the quality of roads in Africa are also lower and poorer than any other region of the world. Alleviating high transportation costs can be achieved by indirect policy interventions and by investment in roads. Indirect policy interventions include market liberalization to reduce local monopolies, removal of road taxes[31] and removing high import tariffs on transportation equipment on spare parts. Improving the quantity and quality of rural infrastructure will require substantial investments. One effective way of doing this, as discussed above, could be the decentralization of fiscal revenues.

Fostering Public and Private Sector Partnerships

Greater collaboration and understanding needs to be developed between the public and private sector. The roles of each sector need to be clearly defined. Governments can improve private sector

---

[31] Not to be confused with user fees.

activity by providing a range of public goods. These include investments in infrastructure (roads and electricity), ensuring rule of law and defending open entry and competition.

Investments in infrastructure such as roads (both in terms of quantity and quality) could improve market access and reduce transaction costs, providing electricity to rural areas may encourage private activity such as private millers and processors, as a common constraint has been the lack of access to energy sources. Subsidies on electricity may be needed, and if they are provided they should be explicitly included in the budget and not hidden in losses of the distribution authority.

Providing key market information (prices, volumes produced, etc.), and ensuring its widespread dissemination will encourage a more rapid response from the private sector. Assuring quality control is also another critical role that requires public-private partnership.

Aid and Input Markets

This study has only addressed one strategic agricultural input, namely fertilizer. In this market across Africa, private sector entry has been limited and volumes traded are small. As previously highlighted non-tariff barriers continue to restrict trade and private sector activity. More significantly, fertilizer aid, in its present form, has caused significant disruptions to fertilizer markets in Africa. A widely recommended strategy is that fertilizer aid should be untied from the purchase of particular types of fertilizer from specific sources and should be provided in cash rather than kind (Donovan and Casey, 1996; Ndayisenga and Schuh, 1995), if this cannot be done, fertilizer aid should be abolished.

As these corrective actions are yet to fully take place, an immediate strategy for African countries should be to improve the effectiveness of the aid and in the specific case of 2KR: there needs to be greater transparency and consistency in the preparation of the requirement list (request letter); an increase in the speed of completing relevant documentation (requests letters, exchange of notes, banking agreement and authorization to pay); the sale of the 2KR aid (fertilizers) in the recipient country at market prices; an elimination of subsidies on transportation and storage; identification of clearly defined uses of the counter part fund; and removal of uncertainty on government actions.

AREAS FOR FURTHER WORK

The complexities and inter-linkages within agriculture make it extremely difficult to cover all aspects of agricultural incentives in one study. While providing some insight into the macroeconomic environment, trade and market access and several domestic price and non-price factors these are by no means the only issues. Further work on incentives needs to examine issues of agricultural labor and gender inequality (Blackden and Bhanu, 1999), rural finance, tenure security, agricultural technology and public rural investment.

Agricultural labor and gender inequality: Where rural food markets are non-existent or at early stages of development, rural households allocate their household labor to ensure domestic food security. In Africa, women account for a larger proportion of the time spent on food production than men. Given the multitude of tasks women undertake in the household (collecting firewood and water, cooking, etc.) and the significant labor constraint they face, more work needs to be done on identifying and developing appropriate labor saving technologies which cover the full range of their domestic and economic tasks. It is only when their labor constraint is reduced will they be more responsive to economic incentives.

Tenure security: Different types of land tenure system have differing impacts on the investment incentives of farmers. In Africa, these systems range from communal tenure through community tenure to individual private tenure.

Rural finance: Lack of access to credit has been highlighted as a key constraint to farmer investments. However, credit alone will not finance the constant and prolonged stream of investments required to capitalize African farms. It is likely that the bulk of farmers in Africa will have to finance most of their investments out of savings in cash or kind. Nevertheless, well functioning financial systems could play a very useful role in the management of savings and in alleviating liquidity constraints. More work is required to develop innovative ways to establish these well functioning financial system.

Agricultural technology: As briefly mentioned in the study the development of new technologies is a key to improving the profitability and agricultural production incentives of African agriculture and maintaining competitiveness on world markets. There has been a significant amount of work and investment into agricultural research and extension, yet land productivity remains virtually stagnant with some improvements in labor productivity. There needs to be further work on developing appropriate technologies and identifying and developing the institutions which need to be in place to ensure adoption of these technologies. Africa will also face the challenge of dealing with biotechnology and intellectual property rights. More work is certainly required to identify the actions needed by African countries in this regard.

Public expenditure: As argued in this study, agriculture has been heavily taxed, yet there has not been a continual period of significant public investment in rural areas. Poor rural infrastructure has raised transaction costs for farmers and inhibited incentives. The range of investment options in rural areas are huge, public investments usually include transportation infrastructure, water supply and irrigation, electrification and communication, agricultural research and extension and investment in human capital. More work is needed to analyze the appropriate levels of public investment in rural areas, the allocation of these investments, their effectiveness and to identify innovative ways of financing these investments.

PART V

COUNTRY PROFILES

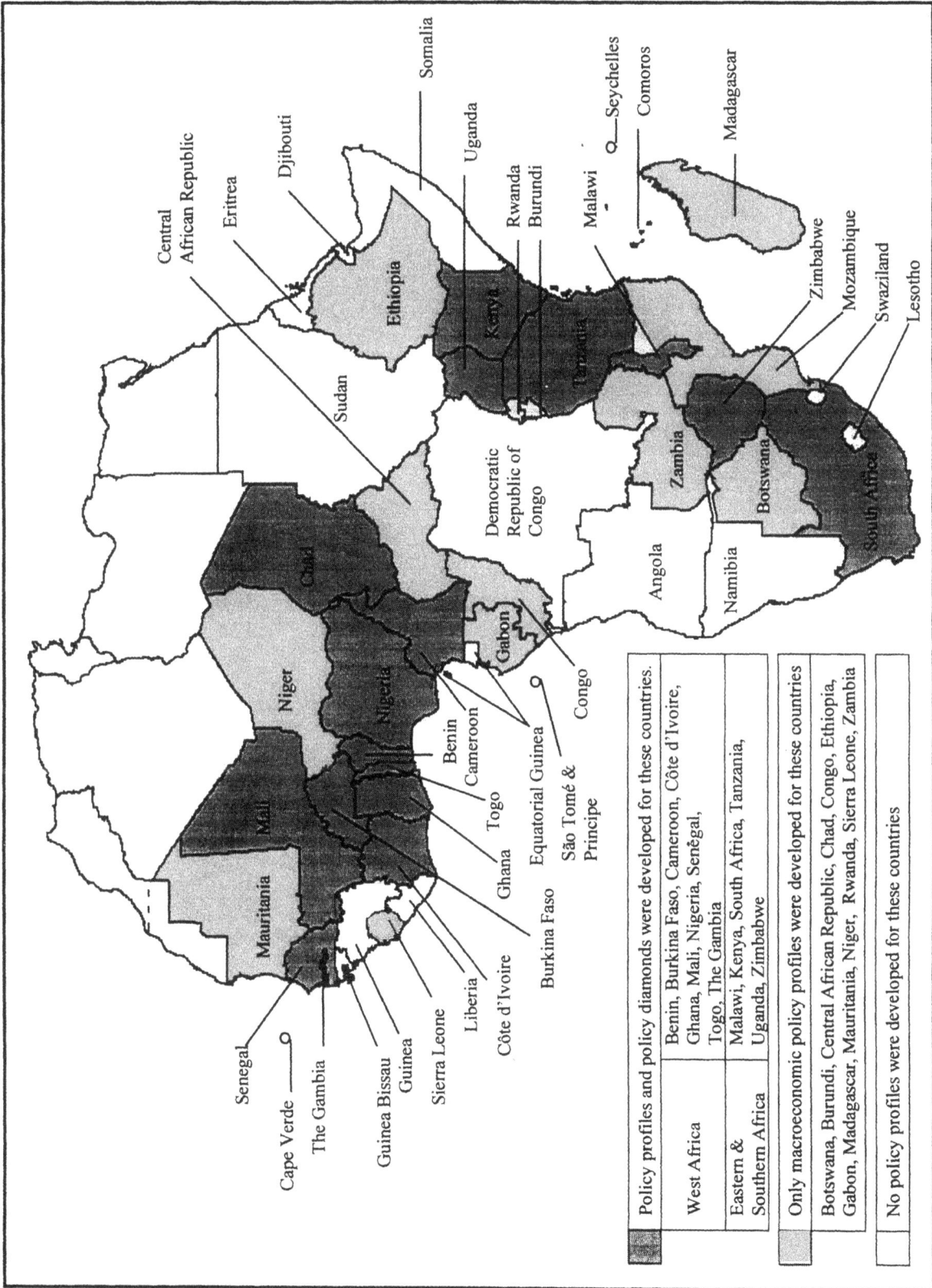

Figure 8.1: Extent of Sub-Saharan African Country Profiles.

Policy profiles and policy diamonds were developed for these countries.

| West Africa | Benin, Burkina Faso, Cameroon, Côte d'Ivoire, Ghana, Mali, Nigeria, Senégal, Togo, The Gambia |
| Eastern & Southern Africa | Malawi, Kenya, South Africa, Tanzania, Uganda, Zimbabwe |

Only macroeconomic policy profiles were developed for these countries

Botswana, Burundi, Central African Republic, Chad, Congo, Ethiopia, Gabon, Madagascar, Mauritania, Niger, Rwanda, Sierra Leone, Zambia

No policy profiles were developed for these countries

DEFINING THE POLICY DIAMONDS

This section attempts to develop pricing policy diamonds for individual African countries (Figure 8.1) Two diamonds are developed, the outer diamond (along the axis) represents the frontier, while the inner diamond represents the current policy stance of the country, measured by some proximate policy indicators. Four policy scores are developed (macroeconomic, export crop, food crop and fertilizer policy) which are represented along each axis of the diamond and are constructed using a blend of price ratios and qualitative assessments. The distance between the two diamonds, along the axis, gives an indication of the distance a particular country's policies are from the perceived frontier.

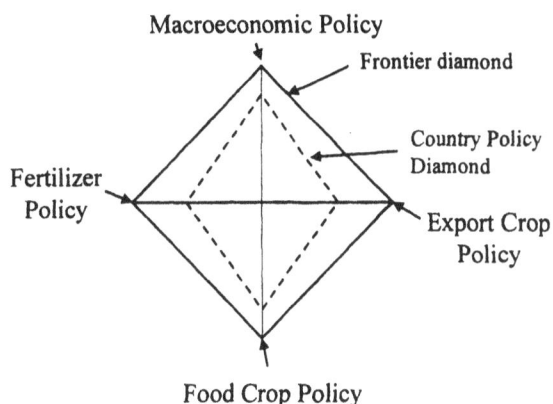

Macroeconomic Policy: The overall macroeconomic policy stance is calculated in the same way as the macroeconomic policy scores in *Adjustment in Africa, 1994* (Table A3 in the appendix). An aggregate score of the fiscal, monetary and exchange rate policy stance is derived using data from 1996-97. These were converted into an index ranging from 0 to 100, where a value of 100 represents the point on the outer diamond. This would be a budget deficit of less that 1.5 percent of GDP, Seigniorage of less that 0.5 percent, Real interest rates of between −3 and 3 and a parallel market premium of less that 10 percent. Export Crops Policy: The producer's share of the f.o.b. price was calculated for the major export crops in the respective countries. A weighted average (by export value) was used to aggregate these shares to form a single share per country. This value was used as the export crop policy score. This indicator not only measures domestic export crop policies (pricing, marketing etc.), but also transportation and transaction costs. However, as the earlier analysis suggested (chapter 5), policy differences explain much of the difference in producer's share of the f.o.b. price. All of the factors which this indicator measures (policy, transaction costs etc) affect agricultural incentives.

Food Crops Policy: This is based on the producer's share of the world price for the major food crop/s in the particular country. In many countries, particularly Southern Africa, this was maize. While it is possible to have a price ratio of greater than one, suggesting a subsidy, all scores were constrained to lie within the frontier (outer diamond). If the ratio was greater than one, the ratio minus one was subtracted from one ie: a 20 percent subsidy would effect the score in the same way as a 20 percent tax. In this sense the analysis focuses both on the farmer and fiscal balance (farmers would nearly always prefer a subsidy to no subsidy), this was only the case for two countries. Qualitative assessments were used to supplement this measure to prevent mis-specification ie: a subsidy, suggesting a ratio of greater than one, may be offset by other policy distortions that decreases the producer price, thus the final score may lie on the frontier. Fertilizer Policy: An aggregate score of trade and pricing policy was constructed using both quantitative data and a qualitative assessment (tariff rates, other trade barriers, pricing policy - Table A9 in the appendix).

AN ANALYSIS OF PRICE POLICIES OF 16 AFRICAN COUNTRIES

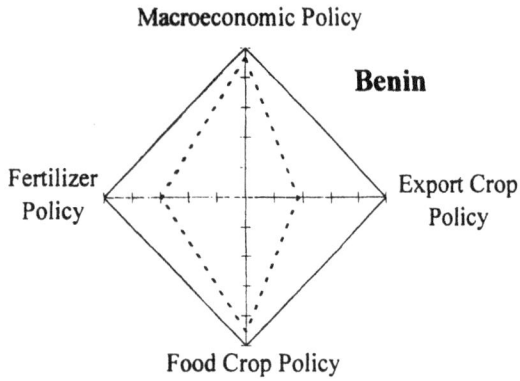

Benin

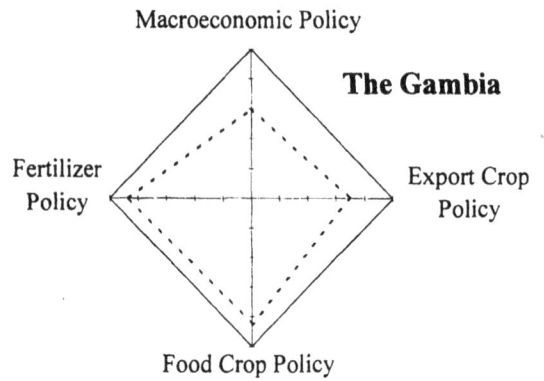

The Gambia

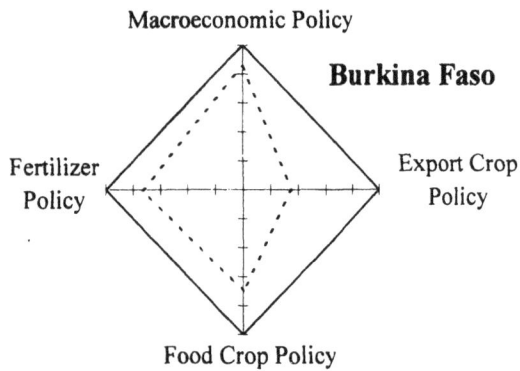

Burkina Faso

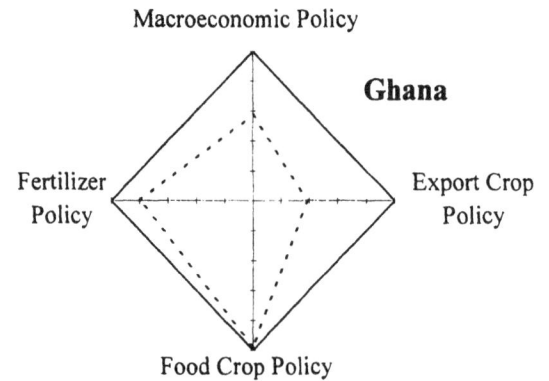

Ghana

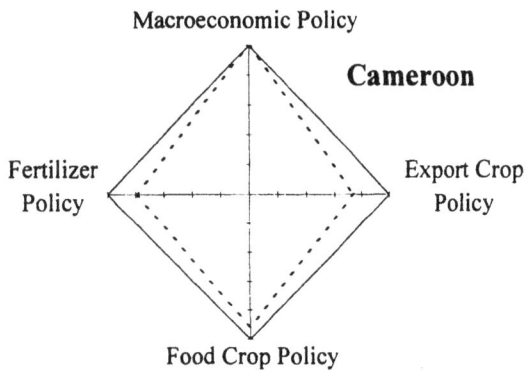

Cameroon

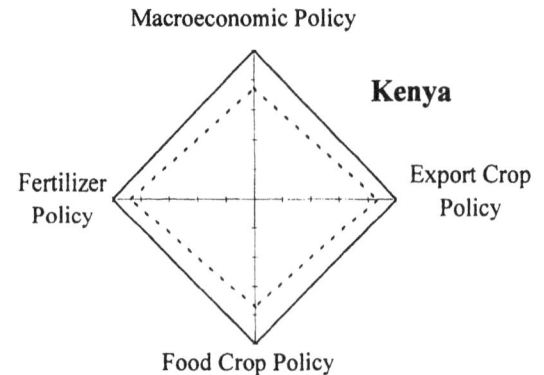

Kenya

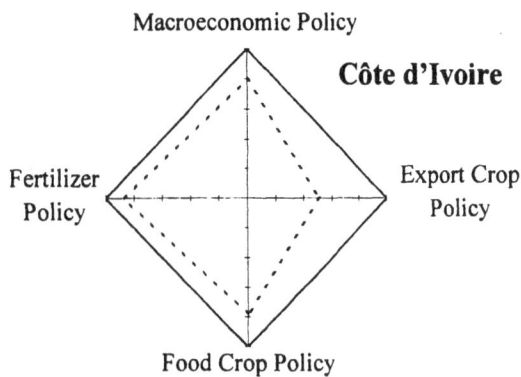

Côte d'Ivoire

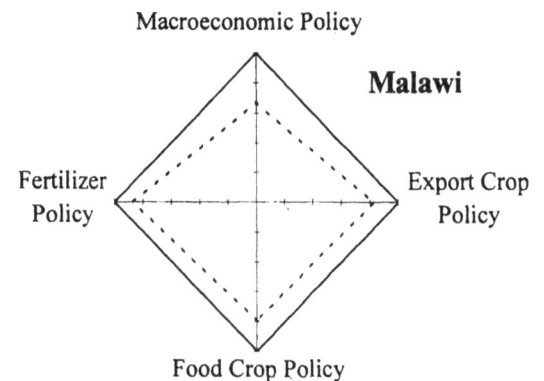

Malawi

POLICY DIAMONDS (*continued*)

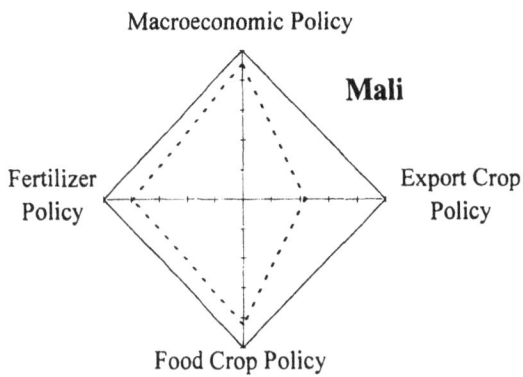

Mali

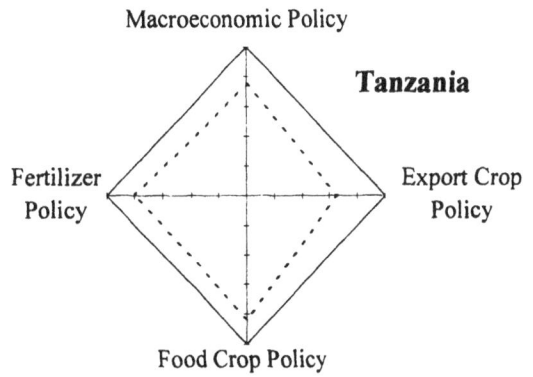

Tanzania

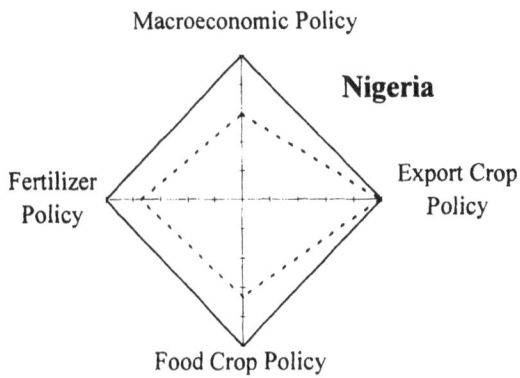

Nigeria

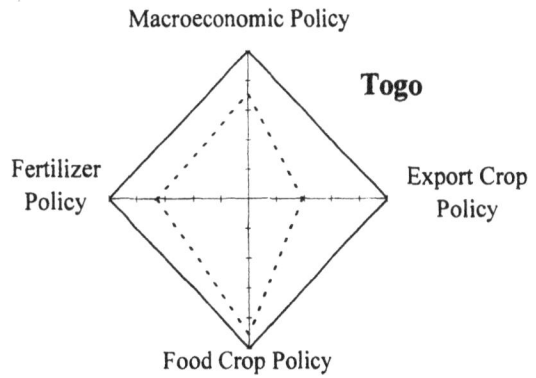

Togo

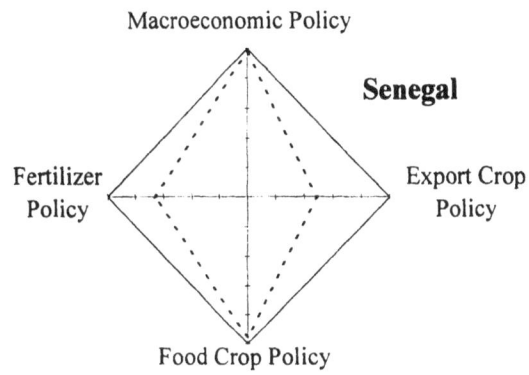

Senegal

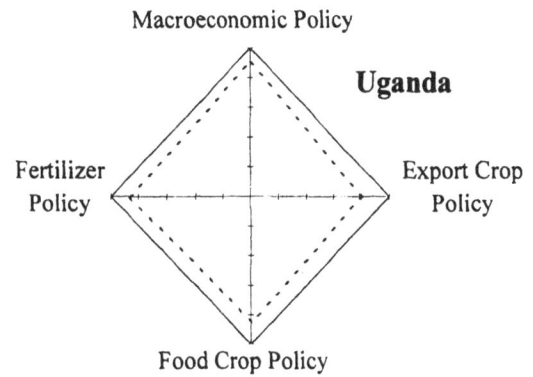

Uganda

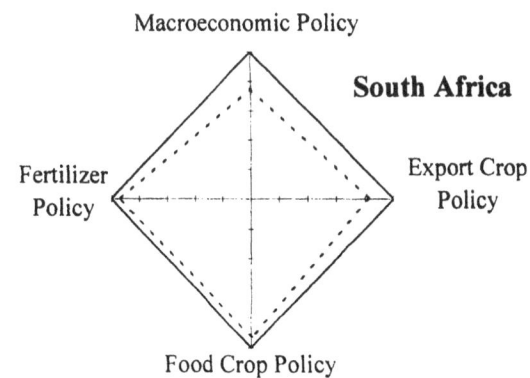

South Africa

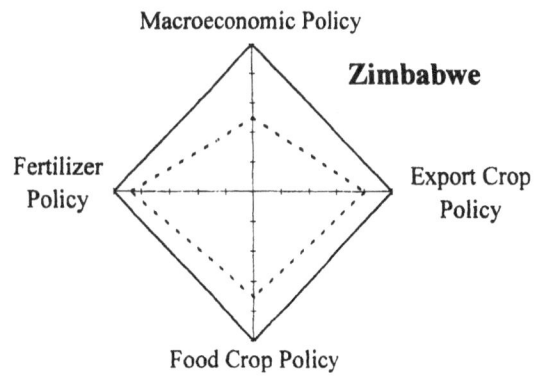

Zimbabwe

137

16 African Countries   **10 West African Countries**   **6 Eastern and Southern African Countries**

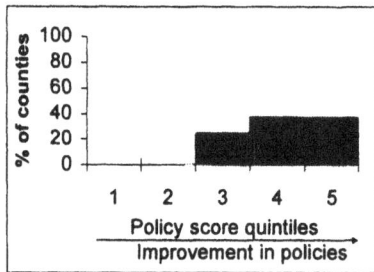
Macroeconomic Policy

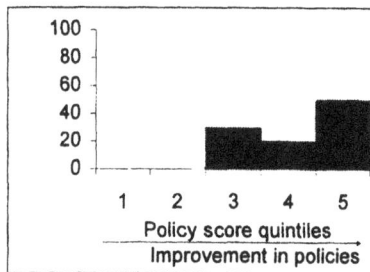
Macroeconomic Policy

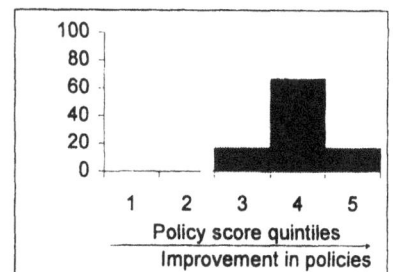
Macroeconomic Policy

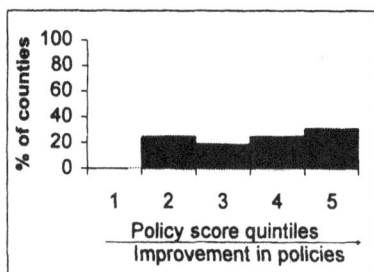
Export Crop Policy

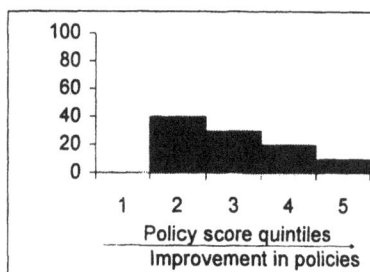
Export Crop Policy

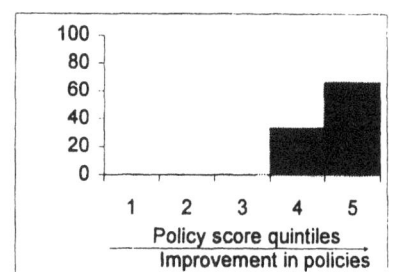
Export Crop Policy

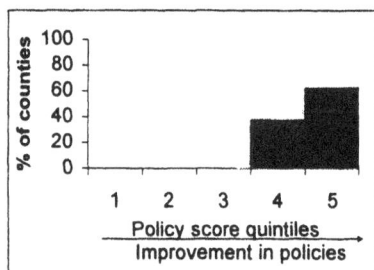
Food Crop Policy

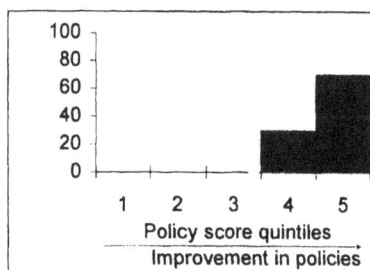
Food Crop Policy

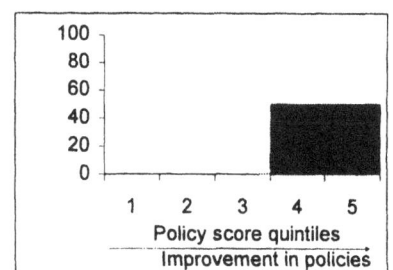
Food Crop Policy

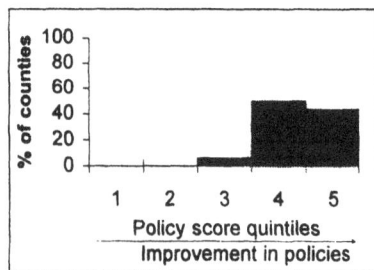
Fertilizer Policy

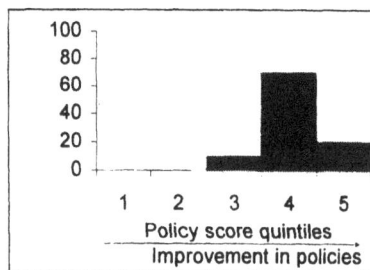
Fertilizer Policy

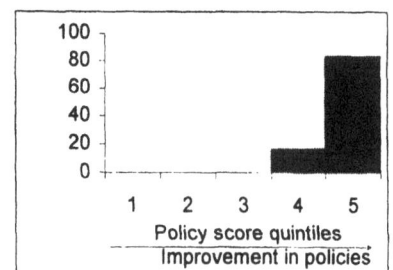
Fertilizer Policy

Figure 8.2: Frequency Distribution of Policy Scores

COUNTRY SUMMARIES

GUIDE TO THE SUMMARY STATISTICS

Macroeconomic indicators: Data for the macroeconomic indicators are from the World Bank and the IMF. See Table A3. The real interest rate was calculated as

$$\text{Real Interest Rate} = \left(\frac{1+r}{1+\rho} - 1\right) * 100$$

where r is the nominal interest rate on deposits and ρ is the inflation rate.

Agricultural Growth: Data used for the agricultural growth indices are from the World Bank and the FAO. The agricultural value added per worker values are from the World Development Indicators, 1998 (see Table A1).

Multi-factor productivity: See *Calculating Agricultural Productivity* on the first page of the appendix and Box A1.

Terms of Trade: see *Calculating the External Trade for Agriculture* in the appendix.
Barter terms of trade: The external barter terms of trade was simply calculated as the ratio of a price index of agricultural commodity exports to a price index of agricultural (food) imports. The indices were constructed using Fischer's ideal index as an aggregate of the major export crops and import crops in the respective countries. The data were obtained from the FAO trade statistics from which a price index was derived as a ratio of value to quantity.
Net income terms of trade: The net income terms of trade was calculated as

$$net\ income\ terms\ of\ trade\ =\ \frac{P_x}{P_m} Q_x$$

where P_x is the agricultural export price index, P_m is the agricultural import price index (in this case food crop imports) and Q_x is the quantity exported. Tsakok, (1990) provides a more extensive definition and highlights some of the limitations of these measures.

Agricultural Exports, Food Crops and Fertilizers: Export and production shares are calculated on the basis of FAO data (See Table A4, A5, A6, A7 in the appendix).

Exchange Rate Pass Through: The exchange rate pass-through was calculated as the percentage change in the real export crop price (adjusted for the trends in the world price) minus the change in the real effective exchange rate. The period over which the change was estimated was 1990 to 1996-97 for export crops. A positive value suggests an enhanced exchange rate pass-through while a negative value suggests that the pass-through has been inhibited (see Table 4.2).

BENIN

General Trends: Benin has enjoyed rapid growth in real agricultural GDP since 1990, increasing at 5.2 percent per annum. Agricultural productivity has also improved with an increase in cereal yields, agricultural value added per worker and multi-factor productivity. Cotton is the dominant export crop accounting for over 90 percent of agricultural exports. Although real cotton producer prices have increased since 1990, with the favorable trends in world prices, extensive intervention in the domestic market has resulted in limited pass-through of the currency devaluation to producer prices, with producers only receiving about 40 percent of the border price.

Pricing Policy Diamond

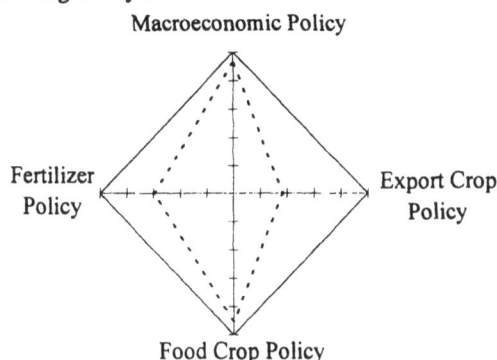

Macroeconomic Policy

Fertilizer Policy

Export Crop Policy

Food Crop Policy

Summary Statistics

Macroeconomic Environment, 1996/97	
Fiscal balance including grants *(% GDP)*	-0.1
Seigniorage *(%)*	0.2
Inflation *(%)*	6.0
Real interest rates *(%)*	1.5
Change in REER since 1990 *(%)*	31
Agriculture's share of GDP *(%)*	38

Agricultural Growth, 1990-1996/97	
Real agricultural GDP growth *(% p.a.)*	5.2
Cereal yields growth *(% p.a. since 1980)*	3.0
Multifactor productivity *(% p.a.)*	1.8
Agric. v.a. per worker (1987 $) *1979-81*	374
Agric. v.a. per worker (1987 $) *1994-96*	563

External Terms of Trade for Agriculture, 1990-1996/97	
Barter terms of trade *(% p.a.)*	-
Net income terms of trade *(% p.a.)*	-
Export price/world fertilizer price *(% p.a)*	-

Agricultural Exports, 1996/97	Cotton
Share of total agricultural exports *(%)*	94
Share of SSA exports *(%)*	11
Annual change in real producer prices *(%)*	2.0
Producers share of f.o.b. price *(%)*	37
Border price/ world price ratio *(%)*	95
Exchange rate pass-through *(%)**	-18

Foods Crops, 1996/97	Cassava	Yams
Share of food crops *(%)*	41	39
Share of SSA food crop production *(%)*	2	4
Annual change in real producer prices *(%)*	-	-

Fertilizer Prices, 1990-1996/97	
Real Urea prices *(% change)*	47
Exchange rate pass-through *(%)*	-

*Price changes and exchange rate pass-through are estimated from 1990 to 1996/97

Macroeconomic Policy: The overall macroeconomic policy stance has been significantly improved. The overall fiscal deficit (including grants) has been reduced and seigniorage and inflation are low. Exchange rate policy has improved with the 1994 devaluation of the CFA franc.

Export Crop Policy: SONAPRA, a government parastatal, dominates the marketing of cotton. Pan-seasonal and pan-territorial prices are announced before planting, Marketing, ginning, transportation and distribution of seed cotton are all controlled by the parastatal.

Food Crop Policy: Since liberalization of maize marketing, the marketing system is completely managed by the private sector. ONASA (Office National d'Appuri a la Securite Altimentaire) is responsible for supporting food security in the country and controls only 0.2 percent of the volume of maize in the market.

Fertilizer Policy: Six companies and their representative distributors served the fertilizer market in 1996/97. The market is heavily regulated by SONAPRA and the Ministry of Rural Development, controlling who can import fertilizer and how much can be imported . The distribution of each company's fertilizer quota is also controlled by designating delivery to specific regions. The government continues to fix the price of fertilizer pan-territorially.

Some remaining policy constraints on prices:
- Pricing policy intervention in the cotton industry continues to provide farmers with a low share of the border price. The pass-though of the currency devaluations to producer prices has been inhibited.
- Fertilizer market regulation and price setting has stifled private sector activity in this market.

BURKINA FASO

General Trends: Since 1990, real agricultural GDP has grown at 3.6 percent per annum. Growth has been particularly strong since 1995, being fueled by the favorable trends in real world commodity prices. Agricultural productivity has shown some improvements with an increase in cereal yields and agricultural value added per worker. Multi-factor productivity, however, has declined. With the increase in the world price of cotton since 1990, the external terms of trade for agriculture have been positive. Real agricultural producer prices have correspondingly increased but the producer's share of the border price remains low and exchange rate pass-through to producer prices has been inhibited.

Pricing Policy Diamond

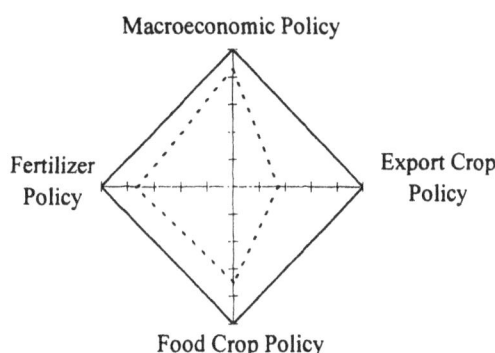

Macroeconomic Policy

Fertilizer Policy

Export Crop Policy

Food Crop Policy

Macroeconomic Policy: Overall improvement since 1990/91. Monetary policy deteriorated slightly with an increase in seigniorage. Fiscal and exchange rate policy has improved significantly with the 50 percent devaluation of the CFA franc in 1994

Export Crop Policy: SOFITEX continues to dominate the cotton industry. This parastatal controls exports, producer prices, transportation and distribution. The price interventions have resulted in farmers receiving a small share of the border price (35 percent) which has declined in recent years.

Food Crop Policy: While there has been progress in liberalization of many food markets, trade restrictions and controls have continued to distort rice and sugar prices.

Fertilizer Policy: Tariff rates on fertilizer remains at 10 percent. SOFITEX dominates the fertilizer market in the cotton growing areas. DIMA distributes aid funded fertilizer to non-cotton regions. Food crop farmers outside the cotton areas have access to fertilizer only when fertilizer aid is available.

Summary Statistics

Macroeconomic Environment, *1996/97*
Fiscal balance including grants *(% GDP)*	-1.9
Seigniorage *(%)*	2.5
Inflation *(%)*	4.2
Real interest rates *(%)*	0.4
Change in REER since 1990 (%)	49
Agriculture's share of GDP *(%)*	35

Agricultural Growth, *1990-1996/97*
Real agricultural GDP growth *(% p.a.)*	3.6
Cereal yields growth *(% p.a. since 1980)*	2.0
Multifactor productivity *(% p.a.)*	-1.4
Agric. v.a. per worker (1987 $) *1979-81*	155
Agric. v.a. per worker (1987 $) *1994-96*	182

External Terms of Trade for Agriculture, *1990-1996/97*
Barter terms of trade *(% p.a.)*	4.2
Net income terms of trade *(% p.a.)*	0.9
Export price/world fertilizer price *(%p.a)*	4.1

Agricultural Exports, *1996/97*
	Cotton
Share of total agricultural exports *(%)*	69
Share of SSA exports *(%)*	7
Annual change in real producer price *(%)*	1.5
Producers share of f.o.b. price *(%)*	35
Border price/ world price ratio *(%)*	96
Exchange rate pass-through *(%)* *	-30

Foods Crops, *1996/97*
	Sorghum	Millet
Share of food crops *(%)*	52	32
Share of SSA food crop production *(%)*	7	6
Annual change in real producer price*(%.)*	-10	-8

Fertilizer Prices, *1990-1996/97*
Urea price *(% change)*	9
Exchange rate pass-through *(%)*	-13

*Price changes and exchange rate pass-through are estimated from 1990 to 1996/97

Some remaining policy constraints on prices:
- Pricing policy intervention in the cotton industry continues to provide farmers with a low share of the border price. The pass-through of currency devaluation to producer prices has been inhibited.
- Price signals in the rice and sugar markets continue to be distorted by trade restrictions and price controls.
- The dominance of SOFITEX in the fertilizer market inhibits the expansion of local private traders who acquire the DIMA supervised fertilizer.

CAMEROON

General Trends: Real agricultural GDP has grown at 4.5 percent per annum since 1990. Favorable world prices for coffee and cotton have improved the external terms of trade for agriculture with a large improvement in the net income and barter terms of trade. Cereal yields have improved but agricultural value added per worker has declined together with multi-factor productivity. Both coffee and cocoa producers have enjoyed a large share of the border price but the exchange rate pass-through to producer prices for cocoa has been inhibited.

Pricing Policy Diamond

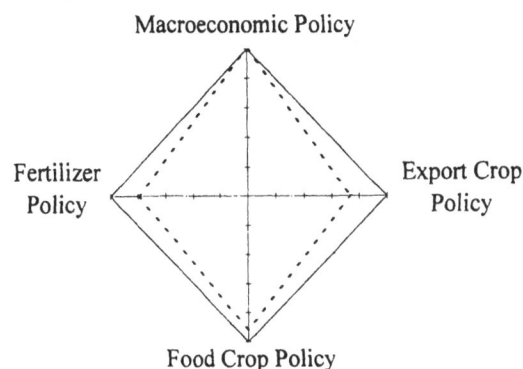

Macroeconomic Policy

Fertilizer Policy

Export Crop Policy

Food Crop Policy

Macroeconomic Policy: Since 1990-91, the overall macroeconomic policy stance has improved significantly with a reduction in the overall fiscal deficit, lower real interest rates and a significant currency devaluation in 1994.

Export Crop Policy: ONCPB (similar to the *caisse de stabilisation*) controlled the marketing of cocoa until 1990. By 1993/94 a more competitive free market and export system had been adopted. Coffee marketing was also controlled by the ONCPB but there has also been a subsequent move to a more competitive market. The cotton sector is gradually being opened up to allow more competition, SODECOTON currently dominates this market.

Food Crop Policy: Ninety five percent of food being marketed is by local private traders, the remainder is marketed by Government institutions and parastatals. The volumes traded are small and the transportation system is poor which has an adverse effect on food distribution efforts.

Fertilizer Policy: All fertilizer subsidies have been removed. Fertilizer import tariffs remain at about 6 percent.

Summary Statistics

Macroeconomic Environment, 1996/97	
Fiscal balance including grants *(% GDP)*	-1.4
Seigniorage *(%)*	-0.9
Inflation *(%)*	5.9
Real interest rates *(%)*	-0.9
Change in REER since 1990 *(%)*	78
Agriculture's share of GDP *(%)*	40

Agricultural Growth, 1990-1996/97	
Real agricultural GDP growth *(% p.a.)*	4.5
Cereal yields growth *(% p.a. since 1980)*	2.2
Multi-factor productivity *(% p.a.)*	-1.1
Agric. v.a. per worker (1987 $) *1979-81*	861
Agric. v.a. per worker (1987 $) *1994-96*	827

External Terms of Trade for Agriculture, 1990-1996/97	
Barter terms of trade *(% p.a.)*	11.9
Net income terms of trade *(% p.a.)*	12.9
Export price/world fertilizer price *(% p.a)*	3.4

Agricultural Exports, 1996/97	Cocoa	Coffee
Share of total agricultural exports *(%)*	25	26
Share of SSA exports *(%)*	6	7
Annual change in real producer prices *(%)*	2.5	9.8
Producers share of f.o.b. price *(%)*	76	73
Border price/ world price ratio *(%)*	86	87
Exchange rate pass-through *(%)* [*]	-37	14

Foods Crops, 1996/97	Cassava	Maize
Share of food crops *(%)*	53	23
Share of SSA food crop production *(%)*	2	2
Annual change in real producer prices *(%)*	-	-1.3

Fertilizer Prices, 1990-1996/97	
Real Urea prices *(% change)*	141
Exchange rate pass-through *(%)*	27

[*]Price changes and exchange rate pass-through are estimated from 1990 to 1996/97

Some remaining policy constraints on prices:
- Interventions continue in the cotton and palm oil sector. SODECOTON and SOCAPALM control cotton and palm oil exports, respectively.
- The inadequacy of transport infrastructure is a significant impediment on development of the food distribution system (inhibiting market integration), for export crops this increases the wedge between the f.o.b. price and the producer price, inhibiting producer incentives.
- High export taxes are still applied to commodity exports.

CÔTE D'IVOIRE

General Trends: Real agricultural GDP has grown at 2.6 percent per annum since 1990. While cereal yields have increased at 1.3 percent per annum, agricultural value added per worker has declined and multi-factor productivity has remained fairly stagnant. The favorable world prices for cocoa and coffee have resulted in favorable external terms of trade. The net income terms of trade have been more favorable than the barter terms of trade reflecting the increase in the volume of agricultural commodities exported. Coffee producer prices have increased significantly but the exchange rate pass-through of the currency devaluation to these prices has been inhibited. The same effect is observed for cotton where the producers share of the border price remains at about 50 percent.

Pricing Policy Diamond

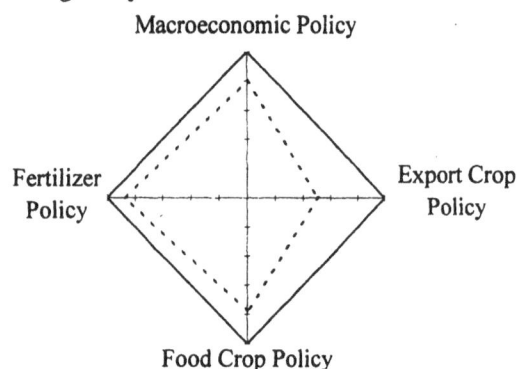

Macroeconomic Policy

Fertilizer Policy

Export Crop Policy

Food Crop Policy

Macroeconomic Policy: Since 1990 there has been a significant improvement in the overall macroeconomic policy stance. The CFA franc was devalued by 50 percent in 1994 which has increased the competitiveness of exports. Inflation and real interest rates are lower and the fiscal deficit has been reduced.

Export Crop Policy: Cocoa and coffee was marketed under a *Caisse de Stabilisation* system. CAISTAB ensured a rigid system of controls and regulations covering the entire marketing chain. Prices were administratively determined and cocoa farmers received 50-55 percent of the f.o.b. price during the 1990s. The cocoa and coffee markets have subsequently (in 1998/99) been liberalized.

Food Crop Policy: Food markets have been liberalized. There has been a recent elimination of transport subsidies and a liberalization of the price controls on all classes of imported rice.

Fertilizer Policy: Fertilizer subsidies have been removed and fertilizer import tariffs were not applied until 1997. Since 1998, the applied rates are 22.6% for mixed fertilizer and 12.5% for simple fertilizer. Exceptions are made for farmers upon request to the Government, in this case the rate is 7.6 %. In the context of UMEOA, new tariffs will apply at a uniform rate of 5%, effective from July 1999.

Summary Statistics

Macroeconomic Environment, 1996/97

Fiscal balance including grants *(% GDP)*	-2.0
Seigniorage *(%)*	1.1
Inflation *(%)*	4.4
Real interest rates *(%)*	1.4
Change in REER since 1980 *(%)*	33
Agriculture's share of GDP *(%)*	28

Agricultural Growth, 1990-1996/97

Real agricultural GDP growth *(% p.a.)*	2.6
Cereal yields growth *(% p.a. since 1980)*	1.3
Multifactor productivity *(% p.a.)*	0.3
Agric. v.a. per worker (1987 $) *1979-81*	1527
Agric. v.a. per worker (1987 $) *1994-96*	1354

External Terms of Trade for Agriculture, 1990-1996/97

Barter terms of trade *(% p.a.)*	3.9
Net income terms of trade *(% p.a.)*	6.8
Export price/world fertilizer price *(% p.a.)*	3.9

Agricultural Exports, 1996/97

	Cocoa	Coffee
Share of total agricultural exports *(%)*	56	16
Share of SSA exports *(%)*	62	18
Annual change in real producer price *(%)*	3.2	9.7
Producers share of f.o.b. price *(%)*	46	72
Border price/ world price ratio *(%)*	96	92
Exchange rate pass-through *(%)* *	-30	-42

Foods Crops, 1996/97

	Yams	Cassava
Share of food crops *(%)*	44	29
Share of SSA food crop production *(%)*	8	2
Annual change in real producer price *(%)*	-	-

Fertilizer Prices, 1990-1996/97

Urea prices *(% change)*	23
Exchange rate pass-through *(%)*	-34

*Price changes and exchange rate pass-through are estimated from 1990 to 1996/97

Some remaining policy constraints on prices:
- Continuing interventions in the cotton market restricts the producers share of the border price and inhibits the pass-through of currency devaluation to producer prices.

THE GAMBIA

General Trends: Although real agricultural GDP has grown at 0.6 percent per annum since 1990, cereal yields, agricultural value added per worker and multi-factor productivity have declined. The dominant export crop is groundnuts with millet and rice being the dominant food crops. The real producer price of groundnuts has decline by -0.1 percent per annum since 1990 and producer's receive about 65 percent of the border price. The real price of rice has increased significantly together with the real fertilizer price.

Pricing Policy Diamond

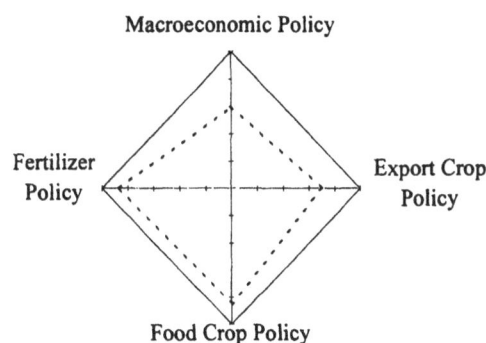

Macroeconomic Policy

Fertilizer Policy

Export Crop Policy

Food Crop Policy

Macroeconomic Policy: The overall macroeconomic policy stance has deteriorated since 1990. The overall fiscal surplus has been eroded with a large current deficit of about 8 percent of GDP. Real interest rates remain high. Exchange rate policy has improved with a reduction of the parallel market exchange rate premium.

Export Crop Policy: Groundnut marketing and exporting have been liberalized and private traders can buy directly from farmers and export groundnuts.

Food Crop Policy: There are no direct controls on producer and consumer prices for sorghum, rice and millet. The government maintains the management of national security stocks.

Fertilizer Policy: Despite liberalized fertilizer markets only a few registered private entrepreneurs have entered the market. Uncertainties in fertilizer demand and in the future role of government and international donors in the fertilizer market have inhibited widespread private sector entry.

Summary Statistics

Macroeconomic Environment, 1996/97	
Fiscal balance including grants *(% GDP)*	-8.2
Seigniorage *(%)*	2.1
Inflation *(%)*	2.5
Real interest rates *(%)*	10.2
Parallel mkt exchange rate premium *(%)*	10.4
Agriculture's share of GDP *(%)*	28

Agricultural Growth, 1990-1996/97	
Real agricultural GDP growth *(% p.a.)*	0.6
Cereal yields growth *(% p.a. since 1980)*	-1.4
Multifactor productivity *(% p.a.)*	-2.6
Agric. v.a. per worker (1987 $) *1979-81*	215
Agric. v.a. per worker (1987 $) *1994-96*	167

External Terms of Trade for Agriculture, 1990-1996/97	
Barter terms of trade *(% p.a.)*	-
Net income terms of trade *(% p.a.)*	-
Export price/World fertilizer price *(% p.a)*	-

Agricultural Exports, 1996/97	G'nuts
Share of total agricultural exports *(%)*	52
Share of SSA exports *(%)*	18
Annual change in real producer prices *(%)*	-0.1
Producers share of f.o.b. price *(%)*	65
Border price/ world price ratio *(%)*	52
Exchange rate pass-through *(%)* [*]	-11

Foods Crops, 1996/97	Millet	Rice
Share of food crops *(%)*	55	17
Share of SSA food crop production *(%)*	0.5	0.2
Annual change in real producer prices *(%)*	-	5.9

Fertilizer Prices, 1990-1996/97	
Real Urea prices *(% change)*	76
Exchange rate pass-through *(%)*	52

[*]Price changes and exchange rate pass-through are estimated from 1990 to 1996/97

Some remaining policy constraints on prices:
- The deterioration of the fiscal balance needs to be reversed - there was some evidence of this in 1997.
- The removal of the uncertainty of government policy is required to encourage private sector entry into product and factor markets.
- Subsidized capital to the current participants in the fertilizer market (the GCU and the NGOs) has inhibited private sector entry. Private traders have limited access to capital to finance private sector fertilizer operations.

GHANA

General Trends: Real agricultural GDP has grown at 2.7 percent per annum since 1990. Multi-factor productivity has also increased together with cereal yields. Although world cocoa prices have increased, the external terms of trade for agriculture have declined as import prices (wheat, rice and sugar) increased at a greater rate than export prices (cocoa). The increased exports of cocoa had some impact at off-setting the decline in net income terms of trade. Cocoa producers share of the border price remains low at about 40 percent

Pricing Policy Diamond

Macroeconomic Policy

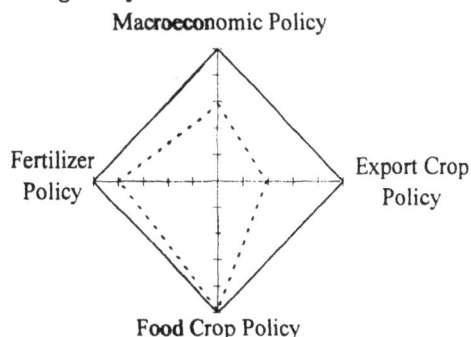

Fertilizer Policy

Export Crop Policy

Food Crop Policy

Macroeconomic Policy: Between 1990 and 1996/97 the overall macroeconomic environment deteriorated significantly. The fiscal deficit increased to about 10 percent of GDP and the inflation rate increased to about 40 percent.

Export Crop Policy: COCOBOD, the cocoa marketing board, controls all stages of cocoa marketing from purchases at the farm-gate through to exports and the sale to domestic processors. A fixed cocoa price is set for the crop year which is less that half of the border price.

Food Crop Policy: Taxes and subsidies have been removed and food crops are traded on open markets. The government abolished the guaranteed minimum price for both maize and rice in 1990. Deteriorating transportation systems have had a negative impact on trade in roots and tubers.

Fertilizer Policy: Although fertilizer subsidies have been removed continuing trade controls (type restrictions and import approval requirements) restrict market entry and competition, contributing to high margins and prices.

Summary Statistics

Macroeconomic Environment, *1996/97*	
Fiscal balance including grants*(% GDP)*	-9.4
Seigniorage *(%)*	2.1
Inflation *(%)*	37.3
Real interest rates *(%)*	-7.6
Parallel mkt exchange rate premium *(%)*	1.6
Agriculture's share of GDP *(%)*	44
Agricultural Growth, *1990-1996/97*	
Real agricultural GDP growth *(% p.a.)*	2.7
Cereal yields growth *(% p.a .since 1980)*	4.5
Multifactor productivity *(% p.a.)*	3.0
Agric. v.a. per worker (1987 $) *1979-81*	813
Agric. v.a. per worker (1987 $) *1994-96*	684
External Terms of Trade for Agriculture, *1990-1996/97*	
Barter terms of trade *(% p.a.)*	-5.9
Net income terms of trade *(% p.a.)*	-2.7
Export price/world fertilizer price *(%p.a)*	-1.3

Agricultural Exports, *1996/97*	Cocoa
Share of total agricultural exports *(%)*	92
Share of SSA exports *(%)*	20
Annual change in real producer price*(%)*	0.7
Producers share of f.o.b. price *(%)*	39
Border price/ world price ratio *(%)*	96
Exchange rate pass-through *(%)* *	-9

Foods Crops, *1996/97*	Yams	Cassava
Share of food crops *(%)*	21	63
Share of SSA food crop production *(%)*	7	8
Annual change in real producer price *(%)*	-	-

Fertilizer Prices, *1990-1996/97*	
Ammonium sulphate prices *(% change)*	46
Exchange rate pass-through *(%)*	47

*Price changes and exchange rate pass-through are estimated from 1990 to 1996/97

Some remaining policy constraints on prices:
- Intervention in cocoa marketing have resulted in a low producers share of the border price.
- The existence of continuing non-tariff barriers increases fertilizer prices. These controls are in the form of limits on the types of fertilizers that can be imported and strict requirements for government approvals on imports.

KENYA

General Trends: Since 1990, real agricultural GDP has grown at 1.0 percent per annum, cereal yields have increased at 0.3 percent together with an increase in multi-factor productivity. Agricultural value added per worker has also shown a marginal decline in the 1990s. The external terms of trade for agriculture has improved significantly, with the net income terms of trade being more favorable than the barter terms of trade. Real producer prices for coffee have been more favorable than for tea, following the trends in world prices. Tea and coffee producers receive about 80 percent of the border price.

Pricing Policy Diamond

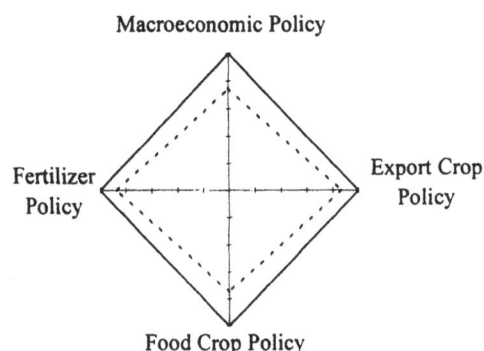

Macroeconomic Policy

Fertilizer Policy

Export Crop Policy

Food Crop Policy

Macroeconomic Policy: The overall fiscal deficit has been significantly reduced from almost 6 percent in 1990 to less than 2 percent in 1996-97. Seigniorage remains high at almost 4 percent. Good exchange rate policy has been maintained with a low parallel market exchange rate premium. The overall macroeconomic policy stance has remained fairly stable since 1990.

Export Crop Policy: Coffee and tea farmers receive over 80 percent of the border price. Small-holder coffee farmers receive lower prices essentially because of deductions made by inefficient co-operatives.

Food Crop Policy: Despite several attempts to reform the maize market, the government continues to intervene. Internal trade has been liberalized but there are continued restrictions on external trade.

Fertilizer Policy: The commercial import of fertilizers has been de-licensed and fertilizer prices throughout the distribution chain decontrolled. Aid-financed fertilizers continue to be allocated in non-transparent ways. Donor financed imports accounted for about 60 percent of annual fertilizer imports. Government owned distributors continue to dominate the market.

Summary Statistics

Macroeconomic Environment, *1996/97*

Fiscal balance including grants *(% GDP)*	-1.6
Seigniorage *(%)*	3.9
Inflation *(%)*	10.1
Real interest rates *(%)*	7.9
Parallel mkt exchange rate premium *(%)*	4.2
Agriculture's share of GDP *(%)*	29

Agricultural Growth, *1990-1996/97*

Real agricultural GDP growth *(% p.a.)*	1.0
Cereal yields growth *(% p.a. since 1980)*	0.3
Multifactor productivity *(% p.a.)*	1.5
Agric. v.a. per worker (1987 $) *1979-81*	268
Agric. v.a. per worker (1987 $) *1994-96*	240

External Terms of Trade for Agriculture, *1990-1996/97*

Barter terms of trade *(% p.a.)*	1.4
Net income terms of trade *(% p.a.)*	3.4
Export price/world fertilizer price*(% p.a)*	1.7

Agricultural Exports, *1996/97*	Tea	Coffee
Share of total agricultural exports *(%)*	33	22
Share of SSA exports *(%)*	74	16
Annual change in real producer prices*(%)*	-6.3	8.2
Producers share of f.o.b. price *(%)*	89	81
Border price/ world price ratio *(%)*	77	95
Exchange rate pass-through *(%)**	1	-8

Foods Crops, *1996/97*	Cassava	Maize
Share of food crops *(%)*	22	64
Share of SSA food crop production *(%)*	1	8
Annual change in real producer prices*(%)*	-	2.2

Fertilizer Prices, *1990-1996/97*	
Real Urea prices *(% change)*	2
Exchange rate pass-through *(%)*	-17

*Price changes and exchange rate pass-through are estimated from 1990 to 1996/97

Some remaining policy constraints on prices:
* Government continues to intervene in the maize markets imposing restrictions on external trade.
* Heavy reliance on fertilizer aid has inhibited the development of private fertilizer traders.
* Deductions made by inefficient co-operatives have resulted in lower prices for small-holder coffee farmers.

MALAWI

General Trends: Since 1990, real agricultural GDP has grown at 6.5 percent per annum. Cereal yields, on average, have declined, together with agricultural value added per worker. Multi-factor productivity growth has remained fairly stagnant. The external barter terms of trade have declined as the export price of tobacco has not increased as much as the price of food imports. However, this decline was almost off-set by the increase in volumes exported. Tobacco producers receive a large share of the export price.

Pricing Policy Diamond

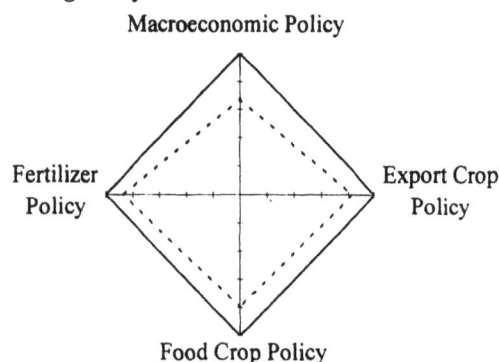

Macroeconomic Policy

Fertilizer Policy

Export Crop Policy

Food Crop Policy

Macroeconomic Policy: Between 1990 and 1996/97, the fiscal deficit increased to 5 percent of GDP. Monetary policy deteriorated with a seigniorage of 5 percent which fueled a 23 percent inflation rate. Exchange rate policy improved with a low parallel market exchange rate premium.

Export Crop Policy: Tobacco production is no longer restricted by quotas, although all marketing takes place through the (privately-run) auctions. The auction sale has been characterized by a low degree of price competition. Marketing restrictions on secondary export crops (legumes, groundnuts and cotton) were removed in the early 1990s. Small-holder tea and coffee producers are no longer required to sell exclusively through crop authorities.

Food Crop Policy: ADMARC continues to sets price bands on maize production although its operations account for less than 15 percent of total production/consumption. Some of ADMARCs holdings have been privatized and maize movement restrictions have been removed.

Fertilizer Policies: Fertilizer subsidies have been removed, and together with the currency devaluation, fertilizer prices have increased significantly. The Small-holder Farmer Fertilizer Revolving Fund of Malawi (SFFRFM) continues to import fertilizer. The commercial market is dominated by a few private sector importer/wholesalers and an expanding network of small stockists.

Summary Statistics

Macroeconomic Environment, 1996/97	
Fiscal balance including grants *(% GDP)*	-5.0
Seigniorage *(%)*	5.0
Inflation *(%)*	23.4
Real interest rates *(%)*	-8.1
Parallel mkt exchange rate premium *(%)*	7.8
Agriculture's share of GDP *(%)*	40

Agricultural Growth, 1990-1996/97	
Real agricultural GDP growth *(% p.a.)*	6.5
Cereal yields growth *(% p.a. since 1980)*	-0.9
Multifactor productivity *(% p.a.)*	-0.1
Agric. v.a. per worker (1987 $) *1979-81*	162
Agric. v.a. per worker (1987 $) *1994-96*	156

External Terms of Trade for Agriculture, 1990-1996/97	
Barter terms of trade *(% p.a.)*	-2.3
Net income terms of trade *(% p.a.)*	-0.8
Export price/world fertilizer price *(%p.a)*	-5.6

Agricultural Exports, 1996/97	Tobacco
Share of total agricultural exports *(%)*	75
Share of SSA exports *(%)*	32
Annual change in real producer prices*(%)*	4.2
Producers share of f.o.b. price *(%)*	82
Border price/ world price ratio *(%)*	86
Exchange rate pass-through *(%)*[*]	54

Foods Crops, 1996/97	Maize
Share of food crops *(%)*	83
Share of SSA food crop production *(%)*	5
Annual change in real producer prices*(%)*	-3.7

Fertilizer Prices, 1990-1996/97	
Real Urea prices *(% change)*	60
Exchange rate pass-through *(%)*	53

[*]Price changes and exchange rate pass-through are estimated from 1990 to 1996/97

Some remaining policy constraints on prices:

- Interventions in the fertilizer market continue to affect commercial prices and the willingness of the private sector to engage in formal import operations and longer term storage.
- There has been some backsliding on earlier reforms in the maize market.
- Internal and external transportation have been identified as causing high fertilizer prices, even after taking into account higher freight and other transportation costs on imported fertilizer.

MALI

General Trends: Since 1990, real agricultural GDP has grown at 3.4 percent per annum. Productivity growth has also shown an improvement with an increase in cereal yields and value added per worker. The external barter terms of trade for agriculture have declined, but an increase in cotton production has resulted in a favorable net income terms of trade since 1990. Currency devaluation has not been fully pass-through to higher producer prices for cotton with producer's receiving about 40 percent of the border price.

Pricing Policy Diamond

Macroeconomic Policy

Fertilizer Policy

Export Crop Policy

Food Crop Policy

Macroeconomic Policy: Overall macroeconomic policy improved significantly between 1990 and 1996/97. The overall fiscal deficit (including grants) declined from about 5 percent to less than 2 percent. Inflation remains low and the CFA franc has been significantly devalued. This significantly improved the competitiveness of Mali exports on international markets.

Export Crop Policy: CMDT continues to control the cotton market. Producer prices are set pan-seasonal and pan-territorial, announced before planting. CMDT is the sole buyer of seed cotton controlling transportation of cotton and inputs required for its production. All cotton gins are owned and operated by the parastatal.

Food Crop Policy: In 1987, there was full consumer and producer market liberalization of coarse grains. Livestock feed is sold under a quota system leading to inefficient distribution. The government maintains strategic stocks of crops for food security reasons.

Fertilizer Policy: CMDT supplies all fertilizer to cotton producers, determining the quantity and type of fertilizer supplied.

Summary Statistics

Macroeconomic Environment, 1996/97	
Fiscal balance including grants *(% GDP)*	-1.5
Seigniorage *(%)*	1.0
Inflation *(%)*	2.9
Real interest rates *(%)*	0.0
Change in REER since 1990 *(%)*	53
Agriculture's share of GDP *(%)*	48

Agricultural Growth, 1990-1996/97	
Real agricultural GDP growth *(% p.a.)*	3.4
Cereal yields growth *(% p.a. since 1980)*	0.8
Multifactor productivity *(% p.a.)*	-1.0
Agric. v.a. per worker (1987 $) *1979-81*	251
Agric. v.a. per worker (1987 $) *1994-96*	259

External Terms of Trade for Agriculture, 1990-1996/97	
Barter terms of trade *(% p.a.)*	-1.4
Net income terms of trade *(% p.a.)*	1.4
Export price/world fertilizer price *(% p.a)*	-3.1

Agricultural Exports, 1996/97	Cotton
Share of total agricultural exports *(%)*	57
Share of SSA exports *(%)*	14
Annual change in real producer prices *(%)*	0.8
Producers share of f.o.b. price *(%)*	44
Border price/ world price ratio *(%)*	82
Exchange rate pass-through *(%)* [*]	-34

Foods Crops, 1996/97	Sorghum	Millet
Share of food crops *(%)*	31	34
Share of SSA food crops production *(%)*	4	6
Annual change in real producer prices *(%)*	-	-

Fertilizer Prices, 1990-1996/97	
Real Urea prices *(% change)*	-
Exchange rate pass-through *(%)*	-

[*]Price changes and exchange rate pass-through are estimated from 1990 to 1996/97

Some remaining policy constraints on prices:
- Pricing policy interventions in the cotton industry continues to provide farmers with a low share of the border price and inhibit the pass-through of currency devaluations to the producer prices.
- The dominance of CMDT in the fertilizer market inhibits the expansion of local private traders
- The quota system for livestock feed leads to inefficient margins.
- Freight transportation is high (from monopolistic tendencies from airlines companies) which impedes the growth in exports of fruit and vegetables.

NIGERIA

General Trends: Between 1990 and 1997, real agricultural GDP has grown at 2.9 percent per annum, aided by the favorable trends in real world commodity prices, particularly for rubber. Although cereal yields have declined, agricultural value added per worker, together with multi-factor productivity, increased. The external terms of trade for agriculture have improved and producer prices for cocoa and rubber have increased significantly. Producers receive a large share of the border price with a complete pass-through of currency devaluations to producer prices.

Pricing Policy Diamond

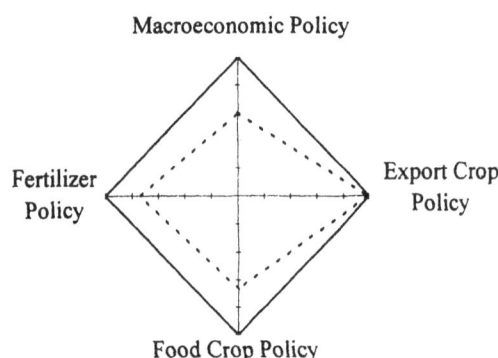

Macroeconomic Policy

Fertilizer Policy

Export Crop Policy

Food Crop Policy

Macroeconomic Policy: Between 1990 and 1997, the overall macroeconomic policy stance deteriorated. While the overall fiscal deficit was small, inflation was high, real interest rates were negative and the parallel market exchange rate premium was large. This decline was largely the result of the collapse of the financial system.

Export Crop Policy: Price controls have been abolished for export crops and marketing boards have been removed (cocoa board). A subsequent surge of private sector activity in the marketing of these crops has resulted in an increase in the producers share of the border price.

Food Crop Policy: A strategic grain reserve and national buffer stock are managed by the government. Import restrictions are imposed on grains such as rice, wheat and maize.

Fertilizer Policy: Subsidies on fertilizers have been significantly reduced. The large differential between the border price and the official price created by past subsidies has induced distributors to smuggle the fertilizer out of Nigeria to be sold on other markets at large profits.

Summary Statistics

Macroeconomic Environment, 1996/97	
Fiscal balance including grants *(% GDP)*	3.3
Seigniorage *(%)*	1.0
Inflation *(%)*	19.9
Real interest rates *(%)*	-12.2
Parallel mkt exchange rate premium *(%)*	274.8
Agriculture's share of GDP *(%)*	43

Agricultural Growth, 1990-1996/97	
Real agricultural GDP growth *(% p.a.)*	2.9
Cereal yields growth *(% p.a. since 1980)*	-1.5
Multifactor productivity *(% p.a.)*	3.6
Agric. v.a. per worker (1987 $) *1979-81*	479
Agric. v.a. per worker (1987 $) *1994-96*	684

External Terms of Trade for Agriculture, 1990-1996/97	
Barter terms of trade *(% p.a.)*	7.0
Net income terms of trade *(% p.a.)*	10.5
Export price/world fertilizer price*(%p.a.)*	9.0

Agricultural Exports, 1996/97	Cocoa	Rubber
Share of total agricultural exports *(%)*	51	24
Share of SSA exports *(%)*	10	32
Annual change in real producer prices*(%)*	6.8	15.2
Producers share of f.o.b. price *(%)*	98	100
Border price/ world price ratio *(%)*	92	93
Exchange rate pass-through *(%)**	73	69

Foods Crops, 1996/97	Yams	Cassava
Share of food crops *(%)*	30	42
Share of SSA food crop production *(%)*	72	37
Annual change in real producer prices*(%)*	-	-

Fertilizer Prices, 1990-1996/97	
Urea prices *(% change)*	8
Exchange rate pass-through *(%)*	35

*Price changes and exchange rate pass-through are estimated from 1990 to 1996/97

Some remaining policy constraints on prices:
- Despite announced policies, the private sector has not been encouraged to participate in the supply of agricultural inputs.
- Government involvement in the food sector inhibits the development of private sector activity in grain markets.
- Import restrictions on grains such as rice, wheat and maize resulted in high domestic food prices.

SENEGAL

General Trends: Real agricultural GDP grew at 2.5 percent per annum between 1990 and 1997. Cereal yields and agricultural value added per worker increased slightly but multifactor productivity has remained fairly stagnant. Real producer prices for cotton and groundnuts have increased but the currency devaluations have not been fully passed-through to these prices. These producers receive about 50 percent of the border price.

Pricing Policy Diamond

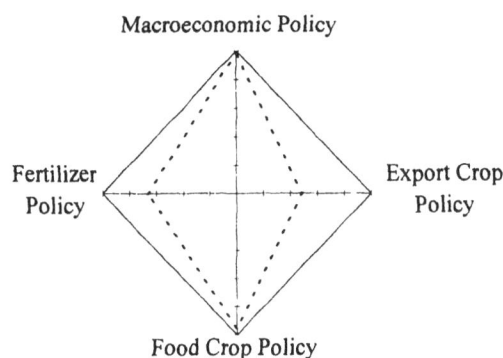

Macroeconomic Policy

Fertilizer Policy

Export Crop Policy

Food Crop Policy

Summary Statistics

Macroeconomic Environment, 1996/97	
Fiscal balance including grants *(% GDP)*	-0.2
Seigniorage *(%)*	0.3
Inflation *(%)*	2.3
Real interest rates *(%)*	3.6
Change in REER since 1990 *(%)*	43
Agriculture's share of GDP *(%)*	18

Agricultural Growth, 1990-1996/97	
Real agricultural GDP growth *(% p.a.)*	2.5
Cereal yields growth *(% p.a. since 1980)*	1.0
Multifactor productivity *(% p.a.)*	-0.1
Agric. v.a. per worker (1987 $) *1979-81*	328
Agric. v.a. per worker (1987 $) *1994-96*	357

External Terms of Trade for Agriculture, 1990-1996/97	
Barter terms of trade *(% p.a.)*	-
Net income terms of trade *(% p.a.)*	-
Export price/world fertilizer price *(% p.a)*	-

Agricultural Exports, 1996/97	G'nuts	Cotton
Share of total agricultural exports *(%)*	4	23
Share of SSA exports *(%)*	9	3
Annual change in real producer prices *(%)*	4.0	3.2
Producers share of f.o.b. price *(%)*	62	47
Border price/ world price ratio *(%)*	75	97
Exchange rate pass-through *(%)* [*]	-42	-12

Foods Crops, 1996/97	Millet	Rice
Share of food crops *(%)*	60	14
Share of SSA food cropw production *(%)*	5	1
Annual change in real producer prices *(%)*	-	-

Fertilizer Prices, 1990-1996/97	
Real Urea prices *(% change)*	-
Exchange rate pass-through *(%)*	-

[*]Price changes and exchange rate pass-through are estimated from 1990 to 1996/97

Macroeconomic Policy: The fiscal deficit remains low. Monetary policy has improved with lower real interest rates and low inflation and seigniorage. The CFA franc was devalued by 50 percent in 1994.

Export Crop Policy: The groundnut market has undergone partial reforms. The state, through the parastatals SONACOS and SONAGRAINES, still control a large part of the marketing and processing system. Pan-territorial and pan-seasonal prices are set. Movement controls on commodities have not been totally removed. SODEFITEX continues to control all aspects of the cotton industry.

Food Crop Policy: Domestic food markets have been recently liberalized. Private wholesalers and retailers have taken over the marketing of rice previously controlled by SAED. Controls on rice imports have also been removed with the dismantling of CPSP.

Fertilizer Policy: Fertilizer subsidies have been eliminated and the private sector has been allowed to enter the market. Although there are a number of private independent traders, SENCHIM continues to dominates the distribution of fertilizer.

Some remaining policy constraints on prices:
- High import tariffs continue to be imposed on raw and refined oil. This import structure has made competition between SONACOS and the private groundnut oil processing sector very difficult. SONACOS continues to dominate the market.
- Cotton farmers continue to receive a low share of the border price and exchange rate devaluations have not been fully transmitted to farmers as higher prices. SÓDEFITEX has been partly restructured to a more commercial entity but continues to control all aspects of the market.

SOUTH AFRICA

General Trends: Real agricultural GDP has grown at 2.5 percent per annum since 1990. Agricultural value added per worker and cereal yields has improved which contributed to the increase in multifactor productivity. The external terms of trade have been more favorable with the upward trends in real world commodity prices. Although positive, the net income terms of trade was not as favorable as the barter terms of trade. Maize accounts for a large share of agricultural exports, but despite favorable world price trends, the removal of the marketing board and price subsidy have resulted in real maize price declines to lower border parity levels.

Pricing Policy Diamond

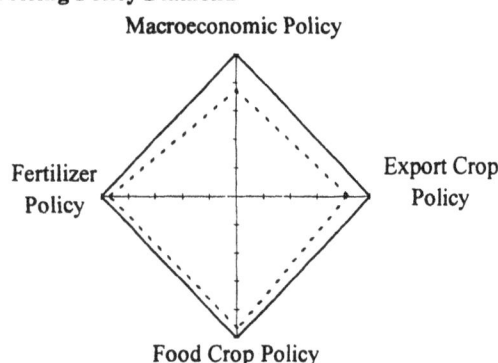

Macroeconomic Policy: Overall macroeconomic policy has been fairly stable since 1990. A fiscal deficit of over 5 percent has been maintained. Inflation has been reduced from double digit to single digit levels. The financial rand has been abolished and the parallel exchange rate market premium is less than 5 percent.

Export Crop Policy: All the marketing boards (dealing with maize, sorghum, oilseeds, wool, meat, cotton, mohair, lucerne, citrus, deciduous fruit, dried fruit, milk and canned fruit) have been closed and producer prices are market determined. Since 1995, a futures market for agricultural commodities has developed. Deciduous and citrus fruit account for almost 30 percent of agricultural exports while sugar and maize account for about 5 and 10 percent respectively.

Food Crop Policy: Wheat and maize are the major food crops. Both of these markets have been liberalized with the removal of marketing boards. Prices are market determined.

Fertilizer Policy: The industry has a high degree of concentration and even though import tariffs are low, domestic prices are much higher than border parity levels.

Summary Statistics

Macroeconomic Environment, 1996/97	
Fiscal balance including grants *(% GDP)*	-5.7
Seigniorage *(%)*	1.2
Inflation *(%)*	8.0
Real interest rates (%)	7.0
Parallel mkt exchange rate premium (%)	3.3
Agriculture's share of GDP *(%)*	5

Agricultural Growth, 1990-1996/97	
Real agricultural GDP growth *(% p.a)*	2.5
Cereal yields growth *(% p.a. since 1980)*	0.4
Multifactor productivity *(% p.a.)*	1.1
Agric. v.a. per worker (1987 $) *1979-81*	2361
Agric. v.a. per worker (1987 $) *1994-96*	2870

External Terms of Trade for Agriculture, 1990-1996/97	
Barter terms of trade *(% p.a.)*	4.6
Net income terms of trade *(% p.a.)*	0.7
Export price/world fertilizer price*(% p.a)*	4.5

Agricultural Exports, 1996/97	Oranges	Maize
Share of total agricultural exports *(%)*	7	10
Share of SSA exports *(%)*	92	64
Annual change in real producer prices*(%)*	-4.9	-1.5
Producers share of f.o.b. price *(%)*	70	93
Border price/ world price ratio *(%)*	75	86
Exchange rate pass-through *(%)* *	-15	-24

Foods Crops, 1996/97	Maize	Wheat
Share of food crops *(%)*	74	22
Share of SSA food crop production *(%)*	24	42
Annual change in real producer prices*(%)*	-1.5	

Fertilizer Prices, 1990-1996/97	
Real Urea prices *(% change)*	6
Exchange rate pass-through *(%)*	-5

*Price changes and exchange rate pass-through are estimated from 1990 to 1996/97

Some remaining policy constraints on prices
- Monopolistic tendencies in the fertilizer market needs to be monitored closely.
- Past policy biases isolated small-scale black farmers from large domestic markets. This isolation and lack of infrastructure afforded to them has created high transaction costs. As a result, price transmissions to these farmers are weak and agricultural incentives have been distorted. Continued efforts are needed to integrate these farmers into the domestic economy.

TANZANIA

General Trends: Real agricultural GDP grew at 3.7 percent per year between 1990 and 1996/97. Agricultural productivity growth has been less impressive with stagnant cereal yields and multi-factor productivity. Favorable commodity world prices have improved the external terms of trade for agriculture with the net income terms of trade exceeding the barter terms of trade. The real domestic producer's price of coffee and cotton have increased significantly since 1990, aided by an improvement in the exchange rate pass-through of the currency devaluation.

Pricing Policy Diamond

Macroeconomic Policy

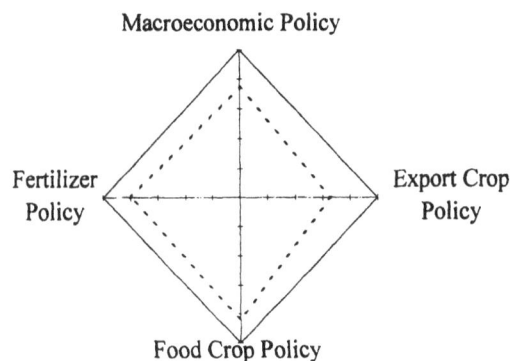

Fertilizer Policy

Export Crop Policy

Food Crop Policy

Macroeconomic Policy: Between 1990 and 1996/97 seigniorage and inflation remained high and negative real interest rates in 1996/97 inhibited improvements in monetary policy. Exchange rate policy improved with significant currency devaluation lowering the parallel market exchange rate premium.

Export Crop Policy: The marketing, processing and pricing of coffee was liberalized in 1990/91. The Tanzanian Coffee Marketing Board (TCMB) has been transformed into the Tanzania Coffee Board which is responsible for policy and regulation of the industry. The role of the cotton marketing board has also been diminished in the processing and sale of this crop. The private sector has entered these markets and has been fairly effective in improving market efficiency. Currency devaluations have been fully transmitted to farmers.

Food Crop Policy: The maize market had been totally liberalized by 1990. Private grain traders handle virtually all of the traded food grains in the country. The government manages a strategic grain reserve to maintain its food security objectives.

Fertilizer Policy: The Tanzania Fertilizer Company (TFC), a government parastatal, continues to import and distribute fertilizer. However, it's market share has been reduced with an increase in private sector activity in fertilizer distribution.

Summary Statistics

Macroeconomic Environment, 1996/97

Fiscal balance including grants *(% GDP)*	-2.1
Seigniorage *(%)*	3.9
Inflation *(%)*	21.4
Real interest rate (%)	-9.6
Parallel mkt exchange rate premium (%)	4.7
Agriculture's share of GDP *(%)*	48

Agricultural Growth, 1990-1996/97

Real agricultural GDP growth *(% p.a.)*	3.7
Cereal yields growth *(% p.a. since 1980)*	0.0
Multi-factor productivity *(% p.a.)*	-1.2
Agric. v.a. per worker (1987 $) *1979-81*	-
Agric. v.a. per worker (1987 $) *1994-96*	-

External Terms of Trade for Agriculture, 1990-1996/97

Barter terms of trade *(% p.a.)*	1.9
Net income terms of trade *(% p.a.)*	8.7
Export price/world fertilizer price*(% p.a)*	3.1

Agricultural Exports, 1996/97

	Coffee	Cotton
Share of total agricultural exports *(%)*	30	29
Share of SSA exports *(%)*	7	10
Annual change in real producer prices*(%)*	14.1	5.9
Producers share of f.o.b. price *(%)*	77	64
Border price/ world price ratio *(%)*	98	83
Exchange rate pass-through (%)*	17	20

Foods Crops, 1996/97

	Cassava	Maize
Share of food crops *(%)*	60	24
Share of SSA food crop production *(%)*	7	7
Annual change in real producer prices*(%)*	-	-0.1

Fertilizer Prices, 1990-1996/97

Real Urea prices *(% change)*	-
Exchange rate pass-through *(%)*	-

*Price changes and exchange rate pass-through are estimated from 1990 to 1996/97

Some remaining policy constraints on prices:

- Numerous local taxes charged by Local Authorities and levies related to crop purchases need to be harmonized.
- Macroeconomic policies need to be stabilized with the reduction in the level of inflation.
- Although tariff levels are relatively low, quality controls of fertilizer imports are virtually absent.

TOGO

General Trends: Real agricultural GDP showed modest growth between 1990 and 1995, but increased significantly in 1996-97. Agricultural value added per worker and cereal yields have increased but multi-factor productivity has remained relatively stagnant. Favorable world cotton prices and the currency devaluation have resulted in an increase in the real domestic producer price of cotton. However, the exchange rate pass-though to producer prices has been inhibited and the producer's share of the border price remains low at about 40 percent.

Pricing Policy Diamond

Macroeconomic Policy

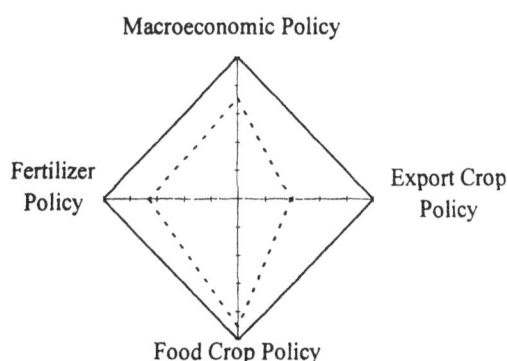

Fertilizer Policy

Export Crop Policy

Food Crop Policy

Macroeconomic Policy: Between 1990 and 1997, the overall macroeconomic policy stance improved. The fiscal deficit remains at about 4 percent of GDP and seigniorage is high. The exchange rate policy stance has improved significantly with the 50 percent devaluation of the CFA franc in 1994.

Export Crop Policy: SOTOCO controls the marketing of cotton and sets pan-seasonal and pan-territorial producer prices. Due to financial difficulties, SOTOCO is being reorganized with greater participation of private firms in the cotton industry. Cotton producers continue to receive a small share (40 percent) of the border price and the pass-through of the currency devaluation to farmer prices has been inhibited.

Food Crop Policy: Food production represents about 70 percent of agricultural GDP. In 1990, all taxes on food crop exports were eliminated. Togograin, the state marketing agency which purchased and distributed crops to smooth out inter-seasonal and inter-regional variation, has been liquidated.

Fertilizer Policy: SOTOCO dominates the distribution of fertilizer to both cotton and maize produces.

Summary Statistics

Macroeconomic Environment, *1996/97*

Fiscal balance including grants *(% GDP)*	-3.5
Seigniorage *(%)*	2.9
Inflation *(%)*	6.4
Real interest rates *(%)*	1.8
Change in REER since 1990 *(%)*	34
Agriculture's share of GDP *(%)*	35

Agricultural Growth, *1990-1996/97*

Real agricultural GDP growth *(% p.a.)*	4.6
Cereal yields growth *(% p.a. since 1980)*	0.7
Multifactor productivity *(% p.a.)*	-0.4
Agric. v.a. per worker (1987 $) *1979-81*	404
Agric. v.a. per worker (1987 $) *1994-96*	461

External Terms of Trade for Agriculture, *1990-1996/97*

Barter terms of trade *(% p.a.)*	-
Net income terms of trade *(% p.a.)*	-
Export price/world fertilizer price *(%p.a)*	-

Agricultural Exports, *1996/97*	Cotton
Share of total agricultural exports *(%)*	60
Share of SSA exports *(%)*	6
Annual change in real producer prices*(%)*	1.4
Producers share of f.o.b. price *(%)*	39
Border price/ world price ratio *(%)*	100
Exchange rate pass-through *(%)**	-27

Foods Crops, *1996/97*	Yams	Cassava
Share of food crops *(%)*	32	32
Share of SSA food crop production *(%)*	2	1
Annual change in real producer prices*(%)*	-	-

Fertilizer Prices, *1990-1996/97*	
Real Urea prices *(% change)*	47
Exchange rate pass-through *(%)*	-8

*Price changes and exchange rate pass-through are estimated from 1990 to 1996/97

Some remaining policy constraints on prices:
- In principle, the fertilizer market is open to private traders, but government-subsidized prices have made it difficult for the private sector to develop.
- Interventions in the marketing of cotton continue to inhibit the pass-through of currency devaluation to producer prices. The result is that producers continue to receive a small share (40 percent) of the producer price.

UGANDA

General Trends: Favorable trends in the real world coffee price have resulted in more favorable external terms of trade for agriculture in Uganda. The improvement in the net income terms of trade have exceeded that of the barter terms of trade. The real producer price of coffee has increased significantly with a full pass-through of the currency devaluation to higher producer prices. The producer's share of the border price has also increased to about 70 percent. These positive trends resulted in an annual growth in real agricultural GDP of 3.8 percent between 1990 and 1997.

Pricing Policy Diamond

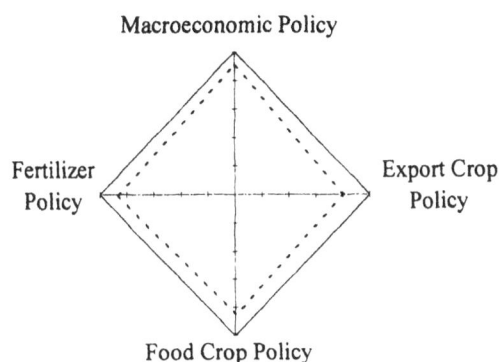

Macroeconomic Policy

Fertilizer Policy — Export Crop Policy

Food Crop Policy

Macroeconomic Policy: The overall macroeconomic policy improved between 1990 and 1996/97 with a lower fiscal deficit, lower inflation and a lower parallel market exchange rate premium. The fiscal deficit remains at about 2 percent of GDP and the inflation rate is about 7 percent.

Export Crop Policy: The Coffee Marketing Board (CMB), a public monopoly, was replaced by the Coffee Marketing Board Limited (CMBL), a parastatal without monopoly powers. The government began issuing export licenses to private firms in 1993, and by mid-1996, the CMBL was handling less than 16 percent of total coffee exports. Cotton marketing and processing has also been liberalized.

Food Crop Policy: Food crops markets have been liberalized and producer prices are market determined. In recent years, maize has become a net export crop.

Fertilizer Policy: The fertilizer markets have been liberalized with a removal of subsidies and government withdrawal from direct procurement and distribution.

Summary Statistics

Macroeconomic Environment, *1996/97*

Fiscal balance including grants *(% GDP)*	-1.9
Seigniorage *(%)*	0.7
Inflation *(%)*	7.0
Real interest rates *(%)*	2.9
Parallel mkt exchange rate premium *(%)*	8.5
Agriculture's share of GDP *(%)*	46

Agricultural Growth, *1990-1996/97*

Real agricultural GDP growth *(% p.a.)*	3.8
Cereal yields growth *(% p.a. since 1980)*	-0.2
Multifactor productivity *(% p.a.)*	2.0
Agric. v.a. per worker (1987 $) *1979-81*	-
Agric. v.a. per worker (1987 $) *1994-96*	592

External Terms of Trade for Agriculture, *1990-1996/97*

Barter terms of trade *(% p.a.)*	8.3
Net income terms of trade *(% p.a.)*	22.2
Export price/world fertilizer price*(% p.a)*	10.9

Agricultural Exports, *1996/97*

	Coffee
Share of total agricultural exports *(%)*	77
Share of SSA exports *(%)*	16
Annual change in real producer prices*(%)*	8.8
Producers share of f.o.b. price *(%)*	72
Border price/ world price ratio *(%)*	78
Exchange rate pass-through *(%)**	19

Foods Crops, *1996/97*

	Cassava	Maize
Share of food crops *(%)*	56	20
Share of SSA food crop production *(%)*	3	2
Annual change in real producer prices*(%)*	-	-

Fertilizer Prices, *1990-1996/97*

Real Urea prices *(% change)*	-
Exchange rate pass-through *(%)*	-

*Price changes and exchange rate pass-through are estimated from 1990 to 1996/97

Some remaining policy constraints on prices:

- Given the importance of coffee production in Uganda and the recent outbreak of coffee wilt disease (*tracheomycosis*), extreme vigilance is required to prevent further infection. Movement restrictions and tree destruction, required to prevent the disease from spreading, disrupt markets and have imposed significant economic shocks on rural households.
- Reliance on fertilizer aid has inhibited private sector development in the fertilizer market. It must be noted that fertilizer use in Uganda is very low.

ZIMBABWE

General Trends: Between 1990 and 1996/97, real agricultural GDP grew at 3.2 percent per annum. Agricultural productivity has been less impressive with a decline in cereal yields, agricultural value added per worker and multi-factor productivity. The barter terms of trade have also been less favorable, however, the net income terms of trade have off-set this decline with an increase in agricultural exports. Real domestic tobacco prices have declined but real cotton prices have increased significantly. Farmers typically receive a large share of the border price with a complete pass-through of the currency devaluation to producer prices.

Pricing Policy Diamond

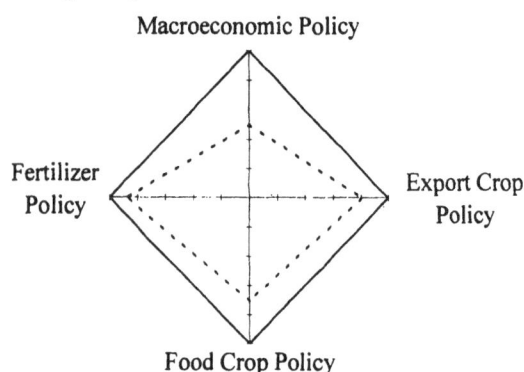

Macroeconomic Policy

Fertilizer Policy

Export Crop Policy

Food Crop Policy

Macroeconomic Policy: Overall macroeconomic policy improved between 1990 and 1996/97. Fiscal deficits, however, remained at close to 10 percent of GDP and the inflation rate was 20 percent. The situation changed dramatically in 1998. The macroeconomic policy stance deteriorated significantly with soaring inflation (>40 percent) and highly volatile exchange rates.

Export Crop Policy: The cotton market has been extensively liberalized, new entrants are permitted and the Cotton Marketing Board has been downsized to become the CCZ (Cotton Company of Zimbabwe) which was privatized in 1998. Producer prices have increased with competition in cotton purchasing. Tobacco is marketed through private auctions and government levies an explicit 5 percent tobacco tax on growers and traders.

Food Crop Policy: Producer price controls, import licensing and foreign exchange controls have been removed. The Grain Marketing Board is a buyer of last resort and has intervened heavily in the market during drought years. There is currently a ceiling price on maize meal.

Fertilizer Policy: Fertilizer producers are protected by a fertilizer import tax which ranges from 0 to 5 percent, and a 10 percent surcharge on the finished product which competes with domestic production. These duties, taxes and tariffs still act as a disincentive to the competitive importation of fertilizer.

Summary Statistics

Macroeconomic Environment, 1996/97	
Fiscal balance including grants *(% GDP)*	-7.6
Seigniorage *(%)*	0.2
Inflation *(%)*	20.4
Real interest rates *(%)*	-1.2
Parallel mkt exchange rate premium *(%)*	7.9
Agriculture's share of GDP *(%)*	14

Agricultural Growth, 1990-1996/97	
Real agricultural GDP growth *(% p.a.)*	3.2
Cereal yields growth *(% p.a. since 1980)*	-2.5
Multifactor productivity *(% p.a.)*	-2.2
Agric. v.a. per worker (1987 $) *1979-81*	294
Agric. v.a. per worker (1987 $) *1994-96*	266

External Terms of Trade for Agriculture, 1990-1996/97	
Barter terms of trade *(% p.a.)*	-1.9
Net income terms of trade *(% p.a.)*	9.6
Export price/world fertilizer price *(% p.a)*	-3.8

Agricultural Exports, 1996/97	Tobacco	Cotton
Share of total agricultural exports *(%)*	60	5
Share of SSA exports *(%)*	60	4
Annual change in real producer prices *(%)*	-2.5	13.8
Producers share of f.o.b. price *(%)*	79	88
Border price/ world price ratio *(%)*	93	99
Exchange rate pass-through *(%)* *	19	111

Foods Crops, 1996/97	Maize
Share of food crops *(%)*	78
Share of SSA food crop production *(%)*	6
Annual change in real producer prices *(%)*	0.6

Fertilizer Prices, 1990-1996/97	
Real Urea prices *(% change)*	22
Exchange rate pass-through *(%)*	-36

* Price changes and exchange rate pass-through are estimated from 1990 to 1996/97

Some remaining policy constraints on prices:

- Macroeconomic stability needs to be restored.
- Mandatory import permits and government approval of fertilizer compositions have negative effects on private sector development in the fertilizer market.
- The Grain Marketing Board sets a ceiling price for maize meal inhibiting private sector activities.

REFERENCES

Adelman, I. (1984). Beyond Export-Led Growth. *World Development*, 12(9):937-49.

Adugna, T. (1997). Factors Influencing the Adoption and Intensity of Use of Fertilizer: The Case of Lume District, Central Ethiopia. *Quarterly Journal of International Agriculture,* 36:173-87.

Agawala, R. (1983). *Price Distortions in Developing Countries.* World Bank, Washington, D.C.

Ahmed, R. and Rustagi, N. (1987). Marketing and Price Incentives in African and Asian Countries: A Comparison, in D. Elz (ed.) *Agricultural Marketing Strategy and Price Policy,* World Bank, Washington, D.C.

Akiyama, T. and Nishio, A. (1997). Indonesia's Cocoa Boom: hands-off policy encouraging small-holder dynamism. *World Bank Policy Research Working paper No. 1600.* World Bank, Washington, D.C.

Ali, A. and Thorbecke, E. (1997). The State of Rural Poverty, Income Distribution and Rural Development in SSA. Paper presented at the *African Economic Research Consortium* on Comparative Development experiences in Asia and Africa, Johannesburg, November 3-6.

Amjadi, A. and Yeats, A. (1995). Have Transport Costs Contributed to the Relative Decline of African Exports?: Some Preliminary Empirical Evidence. *World Bank Policy Research Working Paper, No. 1559.* World Bank, Washington D.C.

Arnade, C.A. (1994). Using Data Envelopment Analysis to Measure International Agricultural Efficiency and Productivity. *United States Department of Agriculture Technical Bulletin No. 1831.* USDA, Washington, D.C.

Badiane, O., Goletti, F., Kherallah, M., Berry, P., Govindan, K., Gruhn, P. and Mendoza, M. (1997). *Agricultural Input and Output Marketing Reforms in African Countries.* International Food Policy Research Institute, Washington, D.C.

Barrett, C.B. (1997). Food Marketing Liberalization and Trader Entry: Evidence from Madagascar. *World Development*, 25(5):763-777.

Barro, R. (1991). Economic Growth in a Cross-Section of Countries. *Quarterly Journal of Economics,* 106(2):407-43.

Barro, R.J. and X. Sala-i-Martin. (1995). *Economic Growth.* McGraw Hill Inc. New York.

Beynon, J., Jones, S. and Yao. S. (1992). Market Reform and Private Trade in Eastern and Southern Africa. *Food Policy,* 17:399-408.

Binswanger, H.P. (1980). Attitudes Towards Risk: Experimental Evidence form Rural India. *American Journal of Agricultural Economics,* 62:964-76.

Binswanger, H.P. (1981). Attitudes towards Risk: Theoretical Implications of Experiment in Rural India. *Economic Journal,* 91, 867-90.

Binswanger, H.P. (1989). *How Agricultural Producers Respond to Prices and Government Investments.* World Bank First Annual Conference on Development Economics. World Bank, Washington, D.C.

Binswanger, H.P., Mundlak, Y., Yang, M.C. and Bowers, A. (1987). On the Determinants of Cross-country Aggregate Agricultural Supply. *Journal of Econometrics*, 36(1):111-131.

Binswanger, H.P. (1992). *Determinants of Agricultural Supply and Adjustment Policies*. Hans Ruthenberg Lecture, Conference of the European Association of Agricultural Economists. September 25. Stuttgart.

Binswanger, H.P. and Pingali, P. (1989). Technical Priorities for Farming in Sub-Saharan Africa. *World Bank Research Observer*, 3, 1, 81-98.

Binswnger, H.P., Townsend, R.F. and Tshibaka, T. (1999). Spurring Agriculture and Rural Development in Africa. Paper presented at the 'Can Africa Claim the 21st Century" Seminar Series. Abidjan, July, 6-10.

Blackden, C.M. and Bhanu, C. (1998). Gender, Growth and Poverty Reduction: Special Program Assistance for Africa, 1998 Status Report on Poverty in Sub-Saharan Africa. *World Bank Technical Paper, No. 428*. World Bank, Washington, D.C.

Blackie, M. J. (1994). *Realizing Small-Holder Agricultural Potential*. In Rukuni, M, and Eicher, K. (eds). Zimbabwe's Agricultural Revolution. University of Zimbabwe Publications.

Bleaney, M. and Greenaway, D. (1993). Long-Run Trends in the Relative Price of Primary Commodities and in the Terms of Trade of Developing Countries. *Oxford Economic Papers*, 45:349-63.

Bloom, D.E. and Sachs, J.D. (1998). Geography, Demography and Economic Growth in Africa. *Brookings Papers on Economic Activity*, 2:207-95.

Bond, M. (1983). Agricultural Responses to Prices in Sub-Saharan Africa. *IMF Staff Papers*, 30:703-26.

Boserup, E. (1965). *The Conditions of Agricultural Growth: The Economics of Agrarian Change Under Population Pressure*, Aldine Publishing Company, New York.

Brunetti, A., Kisunko, G. and Weder, B. (1998). Credibility of Rules and Economic Growth: Evidence From a Worldwide Survey of the Private Sector. *World Bank Economic Review*, 12(3):353-84.

Brüntrup, M. (1997). *Agricultural Price Policy and its Impact on Production, Income, Employment and the Adoption of Innovations: A Farming Systems Based Analysis of Cotton Policy in Northern Benin*. Pert Lang, Frankfurt.

Bumb, B.L. (1988). Fertilizer Supply in Sub-Saharan Africa: An Analysis. In T.B. Tshibaka and C.A. Baanante (eds.), *Fertilizer Policy in Tropical Africa*. Lomé, Togo: International Fertilizer Development Center and International Food Policy Research Institute.

Bumb, B.L., Teboh, J.F., Atta, J.K. and Asenso-Okyere, W.K. (1994). Ghana Policy Environment and Fertilizer Sector Development. *IFDC Technical Bulletin T-41*. Muscle Shoals, Alabama: International Fertilizer Development Center and Ghana Institute of Statistical, Social and Economic Research.

Bureau, J-C, Färe, R. and Grosskopf, S. (1995). A Comparison of Three Nonparametric Measures of Productivity Growth in European and United States Agriculture. *Journal of Agricultural Economics*, 46(3): 309-326.

Byerlee, D. (1991). *Adaptation and Adoption of Seed-Fertilizer Technology: Beyond the Green Revolution*, Presented at the Conference on Mechanisms of Socio-Economic Change in Rural Areas, Canberra, Australia.

Byerlee, D. and Gregory, P. (1999). *Agricultural Biotechnology and Rural Development: Options for the World Bank.* Biotechnology Task Force Discussion Paper, World Bank, Washington, D.C.

Caves, D.W., Christensen, L.R. and Diewert, W.E. (1982). The Economic Theory of Index Numbers and The Measurement of Input, Output and Productivity, *Econometrica* 50(6):1393-1414.

Chakravarti, A. (1997). *Input and Commodity Marketing in Malawi. Status Report.* World Bank Resident Mission, Lilongwe, Malawi.

Chembezi, D. (1989). Estimating Fertilizer Demand and Output Supply for Malawi's Small-Holder Agriculture. *Agricultural Systems.* 33(4): 293-314.

Chhibber, (1989). The Aggregate Supply Response in Agriculture: Survey. In S. Commander (ed.) *Structural Adjustment and Agriculture: Theory and Practice in Africa and Latin America*, Overseas Development Institute, London.

Cleaver, K. (1985). The Impact of Price and Exchange Rate Policies in Agriculture in Sub-Saharan Africa. *World Bank Working Paper No. 728.* World Bank. Washington, D.C.

Collier, P. and Gunning, J.W. (1997). Explaining African Economic Performance. Centre for the Study of African Economies. *Working Paper Series No. 2.1.* Oxford, United Kingdom.

Coetzer, J.A.W., Thomson, G.R. and Tustin, R.C. (eds), *Infectious Diseases of Livestock with Special Reference to Southern Africa.* Oxford University Press. Cape Town.

Conroy, A. (1994). External Transport Costs. *Fertilizer Policy Study: Market Structures, Prices and Fertilizer Use by Small-Holder Maize Farmers.* Harvard Institute for International Development. Economic Planning and Development Department, Office of the President and Cabinet, Government of Malawi.

Conroy, A. (1996*). Fertilizer Use in Malawi: An Overview of Recent Trends, Projection of Effective Demand for the 1996/97 Agricultural Season, Fertilizer Policy Issues and Proposed Monitoring System for the Private Trade.* Discussion paper, Ministry of Economic Planning and Development, Malawi.

Conroy, A. (1997*). Examination of Different Options to Use Fertilizer Subsidies to Promote Maize Productivity.* Discussion paper, Ministry of Finance, Malawi.

Cuddington, J., Urzua, C. and Feyzioglu, T. (1993). *Long-Run Trends in Primary Commodity Prices: Resolving Our Differences Using the ARFIMA Model.* Unpublished memo. Economics Department, Georgetown University, Washington, D.C.

Cukierman, A., Edwards, S. and Tabellini, G. (1992). Seigniorage and Political Instability. *American Economic Review,* 82(3):537-55.

Davenport, M., Hewitt A. and Koning A. (1994). The Impact of the GATT Uruguay Round on ACP States, Overseas Development Institute (London) and European Centre for Development Policy Management (Maastricht).

Deininger, K. and Binswanger, H.P. (1995). Rent Seeking and the Development of Large-Scale Agriculture in Kenya, South Africa and Zimbabwe. *Economic Development and Cultural Change,* 43, 493-522.

Delgado, C. (1992). Why Domestic Food Prices Matter to Growth Strategy in Semi-Open West African Agriculture. *Journal of African Economies*, 1:446-71.

Delgado, C.L., Hopkins, D. and Kelly, V.A. (1998). Agricultural Growth Linkages in Sub-Saharan Africa. *International Food Policy Research Institute Research Report 107*. December.

Demeke, M., Kelly, V., Jayne, T.S., Said, A., Le Vallee, J.C. and Chen, H. (1998). Agricultural Market Performance and Determinants of Fertilizer Use in Ethiopia. *Working Paper 10*, Grain Market Research Project, Ministry of Economic Development and Cooperation, Addis Ababa.

Dercon, S. (1995). On Market Integration and Liberalisation: Method and Application to Ethiopia. *Journal of Development Studies*, 32(1):112-143.

Dercon, S. (1993). Peasant Supply Response and Macroeconomic Policies: Cotton in Tanzania. *Journal of African Economies*, 2(2):157-94.

Donovan, G. (1996). Agriculture and Economic Reform in Sub-Saharan Africa. *AFTES Working Paper, No.18*. Africa Technical Department. World Bank Washington, D.C.

Donovan, G. and Casey, F. (1998). Soil Fertility Management in Sub-Saharan Africa. *World Bank Technical Paper No. 408*. World Bank, Washington, D.C.

Duncan, A. and Jones, S. (1993). Agricultural Marketing and Pricing Reform: A Review of Experience. *World Development*, 21:1495-1514.

Durlauf, S. and Quah, D.T. (1998). The New Empirics of Economic Growth. *National Economic Performance Discussion Paper No. 384*. London School Of Economics. United Kingdom.

Dushmanitch, V.Y. and Darroch, M.A.G. (1990). An Economic Analysis of the Impacts of Monetary Policy on South African Agriculture. *Agrekon*, 29:269-283.

Easterly, W. and Levine, R. (1995). Africa's Growth Tragedy: A Retrospective, 1960-89. *World Bank Policy Research Working Paper*, No. 1503. World Bank. Washington, D.C.

Eicher, C.K. and Staatz, J.M. (1998). *International Agricultural Development*. Third Edition. Johns Hopkins University Press, Baltimore.

Eicher, C.K. (1989). Sustainable Institutions for African Agricultural Development, *ISNAR Working Paper No.19*, International Service for National Agricultural Research, The Hague, Netherlands.

Elbadawi, I. (1992). Real Overvaluation, Terms of Trade Shocks and the Cost to Agriculture in Sub-Saharan Africa. *World Bank Policy Research Working Papers, WPS 831*, World Bank, Washington, D.C.

Fafchamps, M. (1992). Cash Crop Production, Food Price Volatility and Rural Market Integration in the Third World. *American Journal of Agricultural Economics*, 74:90-99.

Fafchamps, M. (1998). *Can Current Theories Explain Africa's Performance?* Department of Economics, Stanford University, Working Paper. http://www.stanford.edu/~fafchamp/engines.pdf.

Färe, R., Grosskopf, S., Norris, M. and Zhang, Z. (1994). Productivity Growth, Technical Progress and Efficiency Change in Industrialized Countries. *American Economic Review*. 84(1):66-83.

Färe, R. Grosskopf, S., Lindgren, B. and Roos, P. (1992). Productivity Changes in Swedish Pharmacies 1980-1989: A Non-parametric Malmquist Approach. *Journal of Productivity Analysis* 3:85-101.

Farrell, M.J. (1957). The Measurement of Productive Efficiency, *Journal of the Royal Statistical Society* A 120, Part 3:253-81.

Fried, H.O., Lovell, C.A.K. and Schmidt, S.S. (1993). *The Measurement of Productive Efficiency: Techniques and Applications*. Oxford University Press, Oxford.

Frisvold, G. and Lomax, E. (1991). Differences in Agricultural Research and Productivity Among 26 Countries, *Agricultural Economic Report no.644*. United States Department of Agriculture, Washington, D.C.

Fulginiti, L.E. and Perrin, R.K. (1998). Agricultural Productivity in Developing Countries. *Agricultural Economics*, 19 (1-2):45-51.

Gardner, B. (1995). *Policy Reform in Agriculture: An Assessment of the Results of Eight Countries*. College Park: University of Maryland.

Gerner, H., Carney, M. and Alognikou, E. (1996). Improving Access to Fertilizer in West Africa Through Pro-Active Fertilizer Privatization Schemes. *African Fertilizer Market*, Vol. 9, No. 1:6:14 (International Fertilizer Development Center).

Gerner, H., Asante, E., Owusu-Bennoah and Marfo K. (1995*). Ghana Fertilizer Privatization Scheme*. Muscle Shoals. IFDC.

Ghura, D. and Just, R.E. (1992). Education, Infrastructure and Instability in East African Agriculture: Implications for Structural Adjustment Programmes. *Journal of African Finance and Economic Development*, Spring, 85-105.

Gisselquist, D. (1998). Pro-Growth Regulatory Reforms for Agricultural Inputs in Malawi, Zambia and Zimbabwe. Memo - World Bank, Washington, D.C.

Gisselquist, D. and Rusike, J. (1998). *Zimbabwe's Agricultural Inputs Industries 1990-96: Regulations, Reforms and Impacts*. Paper presented at the World Bank.

Gisselquist, D. (1996). *A Strategy for Modernizing Agricultural Technology in Ghana*. Paper Presented at Roundtable on: Agricultural Inputs Regulations and Technology Transfer in Ghana, Accra, Ghana.

Gisselquist, D. (1994). Import Barriers for Agricultural Inputs. UNDP-World Bank Trade Expansion Program *Occasional paper No. 10*. New York: United Nations.

Goletti, F. and Babu, S. (1994). Market Liberalisation and Integration of Maize Markets in Malawi. *Agricultural Economics*, 11:311-324.

Government of Zimbabwe (1996). *Zimbabwe's Agricultural Policy Framework, 1995-2020*. Ministry of Agriculture, Zimbabwe.

Green, A.G. and Ng'ong'ola, D.H. (1993). Factors Affecting Fertilizer Adoption in Less Developed Countries: An Application of Multivariate Logistic Analysis in Malawi. *Journal of Agricultural Economics*, 98-109.

Grosskopf, S. (1993). Efficiency and Productivity. In Fried, H.O., Lovell, C.A.K. and Schmidt, S.S. (eds). *The Measurement of Productive Efficiency: Techniques and Applications*. Oxford University Press, Oxford.

Haley, S. and Abbot, P. (1986). Estimation of Agricultural Production Functions on a World-Wide Basis, *Canadian Journal of Agricultural Economics*, 34, 433-54.

Hassan, R. and Hallam, A. (1996). Macro-linkages to Agriculture: A General Equilibrium Model of Sudan, *Journal of Agricultural Economics*, 47(1):66-88.

Hayami, Y. and Ruttan, V. (1971). *Agricultural Development: An International Perspective*. First Ed. Baltimore: Johns Hopkins University Press.

Hayami, Y. and Ruttan, V. (1985). *Agricultural Development: An International Perspective*. Rev. Ed. Baltimore: Johns Hopkins University Press.

Hayami, Y. and Platteau, J-P. (1997). Resource Endowments and Agricultural Development: Africa vs. Asia. Facultes Universitaires Notre-Dame de la Paix, Namur. Faculte des Sciences Economiiques et Sociales. Cashiers: *Serie Recherche (Belgium); No. 192:1-54*.

Heisey, P.W. and Mwangi, W. (1996). Fertilizer Use and Maize Production in Sub-Saharan Africa. *Economics Working Paper 96-01. CIMMYT*.

Hendry, D.F. and Richard, J.F. (1982). On the formulation of Empirical Models in Dynamic Econometrics. *Journal of Econometrics*, 20:3-33.

Herrmann, R. (1997). Agricultural Policies, Macroeconomic Policies and Producer Price Incentives in Developing Countries: Cross-Country Results for Major Crops. *Journal of Developing Areas*, No. 31:203-220.

Hoffmaister, V., Roldos, J.E. and Wickham, P. (1997). *Macroeconomic Fluctuations in Sub-Saharan Africa*. IMF Working Paper, WP/97/82.

IMF. (1998). *The IMF's Response to the Asian Crisis*, Factsheet, April.

Ingco, M., Feder, G. and McCalla, A. (1999). Agriculture and the WTO 2000 Negotiations: Economic Analyses of Issues and Options for Developing Countries. An integrated Program of Research, Policy Analyses and Capacity Building. A Project Proposal to Strengthen Developing Country Participation in the New Multilateral Trade Negotiation. World Bank, Washington, D.C.

Ingco, M. and Townsend, R.F. (1998). *Experience and Lessons from the Implementation of Uruguay Round Commitments: Policy Options and Challenges for African Countries*. Paper for Presentation at the International Workshop on Agricultural Policy of African Countries and Multilateral Trade Negotiations: Challenges and Options, Harare, Zimbabwe, November 23-26, 1998.

Jaeger, W. and Humphreys, C. (1988). The Effect of Policy Reforms on Agricultural Incentives in Sub-Saharan Africa. *American Journal of Agricultural Economics*, 70(5):1036-52.

Jaeger, W.K. (1992). The Effects of Economic Policies on African Agriculture. *World Bank Discussion Papers, Africa Technical Series, No.147*. World Bank. Washington D.C.

Jaffee, S. (1995). Transaction Costs, Risk and the Organization of Private Sector Food Commodity Systems. In Jaffee, S. and Morton, J. (eds) *Marketing Africa's High-Value Foods: Comparative Experiences of an Emergent Private Sector*. Kendall Hunt.

Jaffee, S. and Morton, J. (1995) *Marketing Africa's High-Value Foods: Comparative Experiences of an Emergent Private Sector*. Kendall Hunt.

Jayne, T.S., Khatri, Y. and Thirtle, C. (1994). Determinants of Productivity Change Using a Profit Function: Small-Holder Agriculture in Zimbabwe. *American Journal of Agricultural Economics*, 76:613-18.

Jayne, T.S., Mukumbu, M., Duncan, J., Staatz, J., Howard, J., Lundberg, M., Aldridge, K., Nakaponda, B., Ferris, J., Keita, J. and Sanakoua, K. (1995). Trends in Real Food Prices in Six-Sub-Saharan African Countries. *Policy Synthesis*, USAID, No. 2. (http://www.aec.msu.edu/agecon/fs2/polsyn/no2.htm).

Jayne, T.S. and Jones, S. (1997). Food Marketing and Pricing Policy in Eastern and Southern Africa: A Survey. *World Development*, 25(9)1505-1527.

Jha, D. and Hojjati, B. (1993). Fertilizer Use on Small-Holder Farms in Eastern Province, Zambia. International Food Policy Research Institute. *Research Report No. 94*.

Johnston, B.F. and Mellor, J.W. (1961). The Role of Agriculture in Economic Development. *American Economic Review*, 51(4):566-93.

Jones, S. (1994). *Privatization and Policy Reform: Agricultural Marketing in Africa*. Food Studies Group, University of Oxford, Oxford.

Kelly, V.A. , Crawford, E.W., Howard, J.A., Jayne, T. Staatz, J. and Weber, M.T. (1999). Towards a Strategy for Improving Agricultural Input Markets in Africa. *Policy Synthesis, No. 43*. Global Bureau Office of Agriculture and Food Security. USAID. (http://www.aec.msu.edu/agecon/fs2/polsyn/No43.htm).

Kelly, V., Reardon, T., Tanggen, D. and Naseem, A.L. (1998). Fertilizer in Sub-Saharan Africa: Breaking the Vicious Circle of High Prices and Low Demand. *Policy Synthesis*, No. 32. Office of Sustainable Development, USAID. (http://www.aec.msu.edu/agecon/fs2/polsyn/No32.htm).

Kelly, V., Diagana, B., Reardon, T., Gaye, M. and Crawford, E. (1996). Cash Crop and Foodgrain Productivity: Historical View, New Survey Evidence and Policy Implications. *Policy Synthesis*, No. 7. Office of Sustainable Development, USAID. (http://www.aec.msu.edu/agecon/fs2/polsyn/no7.htm).

Khati, Y., Thirtle, C. and van Zyl, J. (1994). *South African Agricultural Competitiveness: A Profit Function Approach to the Effects of Policy*. In Peter, G.H., Hedley,-Douglas-D., eds. Agricultural competitiveness: Market Forces and Policy Choice: Proceedings of the Twenty-Second International Conference of Agricultural Economists, held at Harare, Zimbabwe 22-29 August 1994. International Association of Agricultural Economists series. Aldershot, U.K.: Dartmouth; distributed by Ashgate, Brookfield, Vt., 1995, pages 670-84.

Kimuyu, P.K., Jama, M.A. and Muturi, W.M. (1991). Determinants of Fertilizer Use on Small-Holder Coffee and Maize in Murang'a District, Kenya. *Eastern Africa Economic Review*, 7:45-49.

Kirsten, J.F. and van Zyl, J. (1996). The Contemporary Agricultural Policy Environment: Undoing the Legacy of the Past. In Van Zyl, J., Kirsten, J. and Binswanger, H.P. (eds), *Agricultural Land Reform in South Africa. Policies, markets and mechanisms*. Oxford University Press. Cape Town.

Krishna, R. (1990). Price and Technology Policies. In *Agricultural Development in The Third World*. 2d ed. Carl K. Eicher and John M. Staatz, eds. (Baltimore: Johns Hopkins University Press), pp. 160-167.

Korani, J. (1992). *The Socialist System*. Princeton: Princeton University Press.

Lamb, R. and Donovan, G. (1992) *Structural Adjustment and African Agriculture: Supply Response to Exchange Rate and Price Movements*, World Bank, unpublished.

Larson, D.F., Varangis, P. and Yabuki, N. (1998). Commodity Risk Management and Development. *World Bank Policy Research Working Paper, No. 1963*. World Bank, Washington D.C.

Ledyard, J.O. (1987). *Market Failure*. Caltech Social Science Working Paper, 623.

Lele, U. and Stone, S.W. (1989) Population Pressure, the Environment and Agricultural Intensification: Variations on the Boserup Hypothesis. *MADIA Discussion Paper No.4*, World Bank, Washington, DC.

Lele, U., Christiansen, and Kadiresan, K. (1989). Fertilizer Policy in Africa: Lessons from Development Programs and Adjustment Lending, 1970-1987. *MADIA Discussion Paper 5*. World Bank. Washington, D.C.

Lewis, P. and Stein, H. (1997). Shifting Fortunes: The Political Economy of Financial Liberalization in Nigeria. *World Development*, 25(1):5-22.

Lipton, M. (1987). Limits of Price Policy for Agriculture: Which Way for the World Bank? *Development Policy Review*, (5):197-215.

Lipton, M. (1991) Market Relaxation and Agricultural Development. In *States and Markets*, (eds). C. Couclough and J. Manor. Clarendon Press. Oxford.

Lloyd, T.A., Morgan, C.W., Ryner, A.J. and Valliant, C. (1997). *The Transmission of World Agricultural Prices in Côte d'Ivoire*. Centre for Research in Economic Development and International Trade (CREDIT), No. 97/15. University of Nottingham.

Maddison, A. (1996). *Monitoring the World Economy: 1820-1992*. Paris: OECD.

Makenete, A.L., Ortmann, G.F. and Darroch, M.G. (1997). Maize Marketing and Pricing in Lesotho: Implications for Policy Reform. *Agrekon*, 36(1):10-26.

Malmquist, S. (1953). Index Numbers and Indifference Surfaces, *Trabajos de Estistica* 4: 209-242.

Mamingi, N. (1996). How Prices and Macroeconomic Policies Affect Agricultural Supply and the Environment. *Policy Research Working Paper 1645*. World Bank, Washington, D.C.

Mbata, J.N. (1997). Factors Influencing Fertilizer Adoption and Rates of Use Among Small-Scale Food Crop Farmers in the Rift Valley of Kenya. *Quarterly Journal of International Agriculture*. 36:285-302.

McCalla, A.F. (1998). *Food Security and the Challenge to Agriculture in the 21st Century*. Paper presented at the Royal Agricultural College Conference, Cirencester, Gloucestershire, England.

McIntire, J. and Varangis, P. (1998). Reforming the Cocoa Marketing and Pricing System in Côte d'Ivoire. *Policy Research Working Paper*, forthcoming. World Bank, Washington, D.C.

McLean, K., Kerr, G. and Williams, M. (1998*). Decentralization and Rural Development: Characterizing Efforts of 19 Countries*. World Bank, Washington, D.C.

Mellor, (1995). *Agriculture on the Road to Industrialization*. International Food Policy Research Institute, Johns Hopkins University Press, Baltimore.

Menon, J. (1995). Exchange Rate Pass-Through. *Journal of Economic Surveys*. 9:197-231.

Meerman, J. (1997). *Reforming Agriculture. The World Bank Goes to Market*. OED Publication. The World Bank, Washington D.C.

Millan, J. and Aldaz, N. (1998). Agricultural Productivity of the Spanish Regions: a Non-Parametric Malmquist Analysis. *Applied Economics*, 30:875-884.

Ministry of Finance and Development Planning. (1997). *National Development Plan 8*, Botswana 1997/98-2002/03.

Morisset, J. (1998). Unfair Trade? The Increasing Gap Between World and Domestic Prices in Commodity Markets During the Past 25 Years. *The World Bank Economic Review*, 12(3):503-26.

Mundlak, Y. and Larson, D.F. (1992). On the Transmission of World Agricultural Prices. *The World Bank Economic Review, 6(3):399-422.*

Ndayisenga, F. and G.E. Schuh. (1995). *Fertilizer Policy in Sub-Saharan Africa*. Sasakawa-Global 2000 Draft Paper. Minneapolis.

Newbery, D. M. G. and Stiglitz, J. E. (1981). *The Theory of Commodity Price Stabilisation: A Study in the Economics of Risk*, Oxford University.

Ng, F. and Yeats, A. (1996). *Open Economies Work Better! Did Africa's Protectionist Policies Cause It's Marginalization in World Trade?* Policy Research Working Paper, No. 1636. World Bank. Washington D.C.

Oyejide, A. (1986). *The Effects of Trade and Exchange Rate Policies on Agriculture in Nigeria*. Research Report No. 55. International Food Policy Research Institute. Washington, D.C.

Oyejide, A. T. (1990). Supply Response in the Context of Structural Adjustment in Sub-Saharan Africa. *AERC Special Paper 1*.

Parker, A.N. (1995). Decentralization: the Way Forward for Rural Development? *World Bank Policy Research Working Paper, No. 1475*. World Bank, Washington, D.C.

Pinstrup-Andersen, P. (1993). *Fertilizer Subsidies: Balancing Short-Term Responses With Long-Term Imperatives*. In N. C. Russell and Doswell, C. Policy Options for Agricultural Development in Sub-Saharan Africa. Centre for Applied Studies in International Negotiations.

Prebisch, R. (1950). *The Economic Development of Latin America and its Principal Problems*, United Nations, New York.

Purcell, G. and Diop, M. (1998). Cotton Policies in Francophone Africa. *forthcoming*. World Bank, Washington, D.C.

Reinhart, C. and Wickham, P. (1994). Commodity Prices: Cyclical Weakness or Secular Decline. *IMF Staff Papers* 41, No. 2:175-213.

Roland-Holst, D.W. and Sancho, F. (1995). Modeling Prices in SAM Structure. *Review of Economics and Statistics*, 77, 361-371.

Rostow, W. (1990). *The Stages of Economic Growth*. Third Edition. Cambridge University Press, New York.

Rukuni, M. and P. Anandajayasekeram (1994), *Getting Agriculture Moving in East and Southern Africa and a Framework for Action*, paper presented at the International Association of Agricultural Economists conference, Harare, Zimbabwe.

Sachs, J.D. and Warner, A.M. (1998). Sources of Slow Growth in African Economies. *Journal of African Economies*, 6(3):335-76.

Sachs, J. (1998). *Address: World Economic Forum*, Southern African Economic Summit, Windhoek, Namibia.

Sahn, D.E., Dorosch, P. and Younger, S. (1996). Exchange Rate, Fiscal and Agricultural Policies in Africa: Does Adjustment hurt the poor? *World Development*, Vol. 24 (4):719-747.

Sánchez, P.A., Valencia, I., Izac, A-M. and Pieri, C. (1995). Soil Fertility Replenishment in Africa. Nairobi, Kenya: ICRAF.

Sapsford, D. and Singer, H. (1998). The IMF, the World Bank and Commodity Prices: A Case of Shifting Sands? *World Development*, 26(9):1661-1675.

Scarborough. V. and Kydd, J. (1992). *Economic Analysis of Agricultural Markets*. Natural Resources Institute, Chatham.

Schiff, M. and Valdes, A. (1992). *The Plundering of Agriculture in Developing Countries*. The World Bank. Washington, D.C.

Schiff, M. and Montenegro, C. (1997). Aggregate Agricultural Supply Response in Developing Countries: A Survey of Selected Issues. *Economic Development and Cultural Change* , 45:393-410.

Schiff, M. and Valdés, A. (1998). Agriculture and the Macroeconomy. *Policy Research Working Paper*. No. 1967. World Bank. Washington, D.C.

Schreiber, G. and Varangis, P. (1999). Cocoa Marketing and Pricing in West Africa. *forthcoming*. World Bank, Washington, D.C.

Schuh, G.E. (1974). The Exchange Rate and U.S. Agriculture. *American Journal of Agricultural Economics, 56:1-13.*

Schuh, E. and G. Norton. (1991). Agricultural Research in an International Policy Context, in: *Agricultural Research Policy, International Quantitative Perspectives*, Pardey, P., Roseboom, J. and J. Anderson (eds), Cambridge University Press, Cambridge.

Schultz, T. (1961). Investment in Human Capital. *American Economic Review*. LI(1):1-17, March.

Shephard, A. and R. Coster. (1987). Fertilizer Marketing Costs and Margins in Developing Countries. Rome: FAO/FLAC (Fertilizer Industry Advisory Council), Memo.

Shephard, R.W. (1953). *Cost and Production Functions*. Princeton, NJ: Princeton University Press.

Singer, H. (1950). The Distribution of Gains Between Investing and Borrowing Countries. *American Economic Review*. Papers and Proceedings, 40, 473-485.

Sodhi, A.J.S. (1999). *The Ethiopian Fertilizer Experience*. Paper presented at the World Bank, Washington, D.C.

Stiglitz, J.E. (1989). Markets, Market Failures and Development. *American Economic Review*, 79 (1):197-210.

Stiglitz, J.E. (1998). *More Instruments and Broader Goals: Moving Toward the Post-Washington Consensus*. The 1998 WIDER Annual Lecture. Helsinki, Finland, January.

Takavarasha, T. (1995). Small-Holder Agricultural Development in Zimbabwe with Particular Reference to the Fertilizer Industry. *FSSA Journal*.

Thirtle, C., Hadley, D. and Townsend, R.F. (1995). Policy Induced Innovation in Sub-Saharan African Agriculture: A Multilateral Malmquist Productivity Index Approach. *Development Policy Review*, 13 (4):323-342.

Timmer, P. (1997). Farmers and Markets: The Political Economy of New Paradigms. *American Journal of Agricultural Economics*, 79:621-627.

Tobin, R. (1996). Pest Management, the Environment, and Japanese Foreign Assistance. *Food Policy*, 21(2):211-228.

Townsend, R.F. (1998). *Macro-economic Policy Effects on Prices and Exports in South African Agriculture*. Department of Agricultural Economics, Extension and Rural Development, University of Pretoria, Memo.

Townsend, R.F. and McDonald, S. (1998). Biased Policies, Agriculture and Income Distribution in South Africa: A Social Accounting Matrix Approach. *Journal of Studies in Economics and Econometrics*, 22 (1):91-114.

Townsend, R.F. and Thirtle. C. (1997). Production Incentives for Small Scale Farmers in Zimbabwe: The Case of Cotton and Maize. *Agrekon*, 36(4):492-500.

Tsakok, I. (1990). *Agricultural Price Policy: A Practitioner's Guide to Partial Equilibrium Analysis*. Cornell University Press.

Tschirley, D.L. and Weber, M.T. (1996) Mozambique Food Security Success Story. *Policy Synthesis*. Number 19. USAID.

Tshibaka, T., Klevor, A.K. and Ntangsi, M. (1998). *Opportunities and Challenges in the Development of Key Sectors of African Economies Under the New WTO Framework*. The Case of Agriculture.

Tshibaka, T. (1993). Agricultural Pricing and the Exchange Rate in Zaire. In Bautista, R, Valdés, A, eds. *The Bias Against Agriculture: Trade and Macroeconomic Policies in Developing Countries*.. San Francisco: ICS Press for the International Center for Economic Growth and the International Food Policy Research Institute. Pg 93-110.

UCDA. (1996). *Annual Report*, 1995/96. Kampala, Uganda.

Valdés, A. (1996). Surveillance of Agricultural Price and Trade Policy in Latin America During Major Policy Reforms. *World Bank Discussion Paper. No. 349*. World Bank. Washington, D.C.

Varangis, P. and D, Larson. (1998). Dealing with Commodity Price Uncertainty. *Policy Research Working Paper*. No. 1667. World Bank. Washington, D.C.

Varangis, P., Akiyama, T. and Mitchell, D. (1995). *Managing Commodity Booms – and Busts*. World Bank, Washington, D.C.

Visker, C., Rutland, D. and Dahoui, K. (1996). *The Quality of Fertilizers in West Africa, 1995*. International Fertilizer Development Centre.

Vogel, S.J. (1994). Structural Changes in Agriculture: Production Linkages and Agricultural Demand-Led Industrialisation, *Oxford Economic Papers*, 46:136-156.

Voortman, R.L., Sonneveld, B.G.J.S. and Keyzer, M.A. (1998). *African Land Ecology: Opportunities and Constraints for Development*. African Economic Research Consortium.

Ward, W., Deren, B. and D'Silva, E. (1990). *A Practitioner's Guide to Shadow Pricing in Agricultural Projects*. World Bank. Washington, D.C.

Weight, D. and Kelly, V. (1998). Restoring Soil Fertility in Sub-Saharan Africa: Technical and Economic Issues. *Policy Synthesis No. 37*. Africa Bureau Office of Sustainable Development, USAID.

Westlake, M. (1999). Malawi: Structure of Fertilizer and Transport Costs. *Technical Report No. 3*. European Food Security Network.

White, H. (1980). A Heteroscedasticity-Consistent Covariance Matrix Estimator and a Direct Test for Heteroscedasticity, *Econometrica*, 48:817-38.

Williams, M. (1992). Balanced Institutional Development: New Concepts of Public and Private Responsibility. *Journal of Society for International Development*, 4:50-53.

World Bank. (1981). *Accelerated Development in Sub-Saharan Africa*. World Bank, Washington, D.C.

World Bank. (1982). *World Development Report 1982*. Oxford University Press. New York.

World Bank. (1989). Improving the Supply of Fertilizers to Developing Countries: A Summary of the World Bank's Experience. Industry and Energy Series. *World Bank Technical paper No. 97*. Washington, D.C.

World Bank. (1994). *Adjustment in Africa. Reforms, Results and the Road Ahead*. Washington, D.C.

World Bank. (1996). *Supply Response: A Review of Experience in Some African Countries*. Global Coalition for Africa Policy Forum, Ouagadougou, Burkina Faso, October 30-31.

World Bank. (1997). *Rural Development: From Vision to Action*. A Sector Strategy Paper. Washington, D.C.

World Bank. (1997). *Mozambique*. Agricultural Sector Memorandum. Volume II: Main Report. Africa Region, World Bank. Washington, D.C.

World Bank. (1997a). Trade and Transport in West and Central African States. Sub-Saharan Africa Transport Policy Program. The World Bank and Economic Commission for Africa. *SSATP Working Paper No. 30.* The World Bank, Washington, D.C.

Data sources

World Bank (1999). World Development Indicators, CD-ROM, World Bank, Washington, D.C.

FAO. (1999). FAOSTAT – Agriculture Data. http://apps.fao.org.

International Monetary Fund. (1998). South Africa - *IMF Staff Country Reports No. 98/96.*

International Monetary Fund. (1998). Benin - Selected Issues and Statistical Appendix. *IMF Staff Country Reports No. 98/88.*

International Monetary Fund. (1998).Nigeria - Selected Issues and Statistical Appendix. *IMF Staff Country Reports No. 98/78.*

International Monetary Fund. (1998). Mauritius - Statistical Annex. *IMF Staff Country Reports No. 98/75* 1998.

International Monetary Fund. (1998). Kenya - Selected Issues and Statistical Appendix. *IMF Staff Country Reports No. 98/66.*

International Monetary Fund. (1998). Burkina Faso - Recent Economic Developments *IMF Staff Country Reports No. 98/65 1998.*

International Monetary Fund. (1998). Uganda - Selected Issues and Statistical Appendix *IMF Staff Country Reports No. 98/61.*

International Monetary Fund. (1998). Republic of Mozambique - Selected Issues *IMF Staff Country Report No. 98/59.*

International Monetary Fund. (1998) Côte d'Ivoire - Selected Issues and Statistical Appendix. *IMF Staff Country Reports No. 98/46.*

International Monetary Fund. (1998) Botswana Selected Issues and Statistical Appendix. *IMF Staff Country Reports No. 98/39.*

International Monetary Fund. (1998) Lesotho - Statistical Annex IMF. *Staff Country Reports No. 98/29.*

International Monetary Fund. (1998) Togo Selected Issues IMF. *Staff Country Reports No. 98/21.*

International Monetary Fund. (1998) Cameroon Statistical Appendix. *IMF Staff Country Reports No. 98/17.*

International Monetary Fund. (1998) Mali - Statistical Annex. *IMF Staff Country Reports No.98/14.*

International Monetary Fund. (1998) Tanzania - Statistical Appendix. *IMF Staff Country Reports No. 98/05.*

International Monetary Fund. (1998) Ghana - Statistical Annex. *IMF Staff Country Reports No. 98/02.*

International Monetary Fund (1991-8). International Financial Statistics.

APPENDIX

Calculating Agricultural Productivity

The techniques applied to calculate productivity indices for countries in SSA have been limited by the lack of input price data. This precludes the cost and profit function approaches and prevents share weighted aggregation which is necessary for an index number approach to constructing multifactor productivity indices. Thus, most existing studies of SSA use production functions with output as the dependent variable (Ghura and Just, 1992; Craig, Pardey and Roseboom, 1994). The Malmquist index provides a solution to some of the data problems. Moreover, it does not require price information for its construction and can provide multilateral comparisons of productivity among countries and over time. The Malmquist has been extensively applied to production data and has several desirable features relative to ideal indices, such as the Tornqvist-Theil and Fisher indices. Rather than assuming that the units of observation are efficient, estimates of efficiency are combined with measures of technical progress. The index is easy to compute, does not require prices and does not rest on behavioral assumptions which is useful if producer's objectives differ, are unknown or not achieved. These properties are particularly advantageous in applications to agriculture in SSA.

The index is based on defining an efficiency frontier which can be expressed in terms of minimising the input requirements per unit of output, following Farrell (1957). The efficient observations define the frontier and the efficiency of all other observations are measured relative to the frontier. Then, the time series dimension of the data allows for estimation of technical progress (the movement of the frontier) and changes in efficiency over time (the distance of the inefficient units from the best practice frontier). This method (see Box A1) has been used to calculate productivity indices for Sub-Saharan African countries (Table A1, MFP [multi-factor productivity column]).

The output used was aggregate agricultural output expressed in 1979/81 'international dollars' (derived using purchasing power parity conversion rates) from the FAO.

The inputs used were arable agricultural land, in thousands of hectares; labor, (economically active population in agriculture), in thousands; fertilizer, tons used and machinery, the number of tractors currently used.

Box A1: The Malmquist Productivity Index.

The non-parametric approach, introduced by Farrell (1957), is used here largely because it does not require prices and leads directly to the Malmquist index. The Farrell technical efficiency measure is defined so that the isoquant, which is the locus of the efficient points that form the boundary of the input requirements set, $L^t(y^t)$, designates the minimal set of inputs, x^t, resulting in the unit level of output of y^t. In this example, the efficiency of the other countries is measured radially relative to this isoquant (Figure 1.3). The figure below explains the Malmquist index, which is based on the Farrell efficiency measure. These measures have been used to calculate indices for firms (Färe *et al* 1992), regions (Millan and Aldaz, 1998) and countries (Färe *et al*, 1994, Bureau *et al*, 1995)

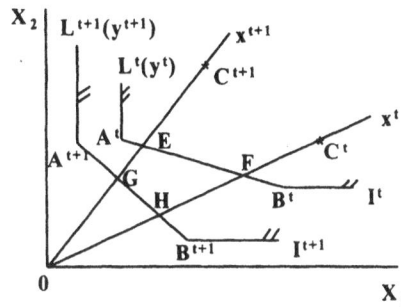

For year t, the efficiency frontier is I^t and for the inefficient country C^t, the Farrell efficiency measure is OF/OC^t. The Malmquist index is defined in terms of Shephard's (1953) distance function, which is simply the inverse of the Farrell measure. The analysis so far is based on cross-sectional estimation of efficiency measures relative to the best-practice frontier for each year. The time series dimension is used to estimate the shifting of the frontier over time, giving a measure of pure technical progress. Thus, inter-temporal and inter-country distance functions form the Malmquist MFP index. Following Färe *et al.* (1992), the Malmquist MFP index is defined, for country *i* and year *t+1*, in terms of distance functions as

$$M_i^{t+1}(y_i^{t+1}, x_i^{t+1}, y_i^t, x_i^t) = \frac{D^{t+1}(y_i^{t+1}, x_i^{t+1})}{D^t(y_i^t, x_i^t)} \left[\frac{D^t(y_i^t, x_i^t)}{D^{t+1}(y_i^t, x_i^t)} \frac{D^t(y_i^{t+1}, x_i^{t+1})}{D^{t+1}(y_i^{t+1}, x_i^{t+1})} \right]^{1/2}$$

The first ratio on the right hand side is the efficiency measure for year t+1 relative to that for year t. These distance functions compare observations for year t+1 with the year t reference technology (isoquant). The other four distance functions define the shift of the technical progress frontier. Two of these terms are identical to the efficiency measures just described. The other two compare year t observations with the (t+1) reference technology, or vice versa.

The base-year frontier, I^t, is defined by two hypothetical efficient agricultural sectors (countries), with observations A^t and B^t, and the efficiency of a third country, at C^t, is calculated relative to this frontier. Thus, $D_c^t(y^t, x^t)$ is equal to OC^t/OF, on the initial vector x^t. In the next year, the frontier has moved to I^{t+1} and country C now has an input ratio x^{t+1} and is at C^{t+1}. The efficiency measure is now OC^{t+1}/OG and the efficiency term in the above equation is measured by these ratios.

$$M_i^{t+1}(y_i^{t+1}, x_i^{t+1}, y_i^t, x_i^t) = \left[\frac{OC^{t+1} / OG}{OC^t / OF} \right] \left[\frac{OH}{OF} \frac{OG}{OE} \right]^{1/2}$$

The first technical progress term is already defined above and is divided by $D^{t+1}(y^t, x^t)$, which is equal to OC^t/OH, giving the ratio OH/OF. The Figure shows that this is the shift of the frontier measured at the factor ratio x^t. Following the same reasoning, the last term in the first equation is measured by OG/OE in the Figure and this is the shift of the frontier measured at the factor ratio of the second period, x^{t+1}. The technical progress element of the Malmquist MFP index is, very reasonably, the geometric mean of the two alternative measures. The methodology of Färe *et al.* (1992), which is described above, is soundly based on index number theory and suits the current problem.

Table A1: Agricultural Output and Productivity Growth Rates for Sub-Saharan African Countries, 1989/90-1996/97.

Region (% of total production in SSA)	Country	Share of Agric. in Region %	Real Agricultural GDP Growth Average Annual % Growth			Popn. Growth Rate	Cereal Yields Kg's per ha.		Growth Rates	Agricultural Value Added Per Worker 1987 US$		Agricultural Value Added Per Hectare of Agricultural Land 1987 US$		MFP Growth Rate
		1997	1970-1980	1980-1990	1990-1997	1990-1997	1979-1981	1995-1997	1980-1997	1979-1981	1994-1996	1979-1981	1992-1994	1980-1996
West Africa	Benin	4	1.8	5.5	5.2	2.9	698	1067	3.0	374	563	188	321	1.8
	Côte d'Ivoire	18	2.7	0.3	2.6	3.0	869	1100	1.3	1527	1354	195	212	0.3
	Ghana	15	-0.3	1.0	2.7	2.7	807	1383	4.5	813	684	215	227	3.0
(28%)	Guinea	3	1.6	1.5	4.4	2.7	958	1251	2.2	-	225	-	54	-
	Guinea- Bissau	1	-1.2	4.7	5.5	2.2	711	1422	4.6	186	292	54	78	-
	Nigeria	55	-0.1	3.3	2.9	3.0	1269	1191	-1.5	479	684	111	150	3.6
	Sierra Leone	2	5.9	3.1	1.6	2.5	1249	1223	-0.6	365	344	-	-	0.8
	Togo	2	1.9	5.6	4.6	3.1	729	830	0.7	404	461	119	189	-0.4
	All countries	**100**	**0.6**	**2.3**	**2.7**	**2.9**	**1096**	**1184**	**0.1**	**-**	**-**	**-**	**-**	**-**
Sahelian Africa	Burkina Faso	5	1.0	3.1	3.6	2.8	575	754	2.0	155	182	64	93	-1.4
	Chad	3	-0.4	2.3	5.4	2.6	587	627	0.7	148	198	6	10	1.2
	The Gambia	0.3	4.8	0.9	0.6	3.6	1284	1035	-1.4	215	167	162	199	-2.6
(27%)	Mali	6	5.0	3.3	3.4	2.8	804	909	0.8	251	259	24	33	-1.0
	Mauritania	2	-1.0	1.7	4.8	2.6	384	744	4.0	301	439	5	7	1.1
	Niger	5	-3.7	1.7	2.2	3.4	440	325	-1.7	292	256	57	63	1.8
	Senegal	5	1.3	2.8	1.6	2.6	690	779	1.0	328	375	92	116	-0.1
	Sudan	75	-	-	-	2.1	645	523	0.0	889	-	42	-	-0.1
	All countries	**100**	**0.4**	**2.6**	**3.1**	**2.6**	**562**	**584**	**0.10**	**-**	**-**	**-**	**-**	**-**
Central Africa	Cameroon	46	3.7	2.1	4.5	2.9	849	1313	2.2	861	827	252	313	-1.1
	Central Afr. Rep	8	1.9	1.6	3.2	2.2	529	971	3.2	456	516	96	119	0.3
	Congo, DRP	4	1.4	2.5	3.1	3.2	807	780	-0.4	218	219	83	113	-1.1
(10%)	Congo, Rep	37	2.5	3.4	0.9	2.9	825	818	0.5	544	629	21	28	0.4
	Gabon	4	1.0	1.2	-2.3	2.6	1718	1754	0.0	1412	1516	67	74	3.3
	All countries	**100**	**2.4**	**2.2**	**3.1**	**3.1**	**804**	**951**	**0.7**	**-**	**-**	**-**	**-**	**-**

Sources: Real agricultural GDP growth and cereal yields are from the 1999 World Development Indicators; population growth and yield growth are calculated from World Bank data, agricultural value added per worker and agricultural value added per hectare of agricultural land are from the 1998 World Development Indicators. Multi-factor productivity (MFP) is calculated from FAO and World Bank Data (see Box A1).

Table A1: Agricultural Output and Productivity Growth Rates for Sub-Saharan African Countries, 1989/90-1996/97 *(continued)*.

Region (% of total production in SSA)	Country	Share of Agric. In Region % 1997	Real Agricultural GDP Growth (Average Annual % growth) 1970-1980	1980-1990	1990-1997	Popn. Growth rate 1990-1997	Cereal Yields (Kg's per ha.) 1979-1981	1995-1997	Growth rates 1985-1995	Agricultural Value Added Per Worker 1987 US$ 1979-1991	1994-1996	Agricultural Value Added Per Hectare of Agricultural Land 1987 US$ 1979-1991	1992-1994	MFP Growth Rate 1980-1996
East Africa (18%)	Burundi	4	2.2	3.1	-3.0	2.6	1081	1378	1.7	218	177	212	270	1.1
	Ethiopia	30	1.2	0.6	3.0	2.1	-	1229	-	-	181	-	116	0.0
	Kenya	18	4.8	3.3	1.0	2.6	1364	1634	0.3	268	240	68	90	1.5
	Madagascar	7	0.4	2.5	1.6	2.8	1664	1992	1.2	190	178	26	34	-0.1
	Mauritius	2	-3.3	2.9	0.4	1.2	2536	4664	3.3	1764	3762	1607	1902	1.3
	Rwanda	4	8.0	0.5	-7.4	0.0	1134	1365	0.6	306	206	445	378	-1.6
	Uganda	36	-2.0	2.1	3.8	3.2	1555	1331	-0.2	-	592	-	515	2.0
	All countries	**100**	**1.0**	**2.2**	**2.0**	**2.3**	**1313**	**1384**	**0.6**	-	-	-	-	-
Southern Africa (16%)	Angola	5	-5.3	0.5	-6.8	3.1	526	542	0.1	-	149	-	9	1.1
	Botswana	1	8.3	2.2	-0.8	2.4	203	290	0.9	392	483	4	5	-1.3
	Namibia	3	-1.2	1.8	4.0	2.7	377	311	-2.5	1295	1458	8	9	-
	Lesotho	1	0.3	2.2	4.0	2.1	977	976	1.2	291	194	35	24	-1.5
	Mozambique	10	-1.7	2.1	5.5	4.3	603	765	1.4	-	92	-	12	0.7
	Malawi	6	4.4	2.0	4.7	2.7	1161	1216	-	162	156	145	153	0.1
	South Africa	46	3.2	2.9	2.5	1.7	1845	1839	0.4	2361	2870	45	49	1.1
	Tanzania	17	3.9	2.1	3.7	3.0	1911	1317	0.0	-	-	-	-	-1.2
	Zambia	2	2.1	3.6	0.7	2.8	1676	1574	-0.6	116	100	6	7	0.4
	Zimbabwe	10	0.6	3.1	3.2	2.4	1359	1095	-2.5	294	266	34	41	0.1
	All countries	**100**	**1.3**	**2.1**	**2.5**	**2.7**	**1392**	**1457**	**0.0**	-	-	-	-	-
SSA	**All regions**		1.1	2.3	2.5	2.7	1089	1050	-	458	392	53	68	-

Sources: Real agricultural GDP growth and cereal yields are from the 1999 World Development Indicators; population growth and yield growth are calculated from World Bank data, agricultural value added per worker and agricultural value added per hectare of agricultural land are from the 1998 World Development Indicators. Multifactor productivity (MFP) is calculated from FAO and World Bank Data (see Boa A1).

Sub-Saharan African countries not included in the table include: Comoros, Equatorial Guinea, Sao Tome, Seychelles, Swaziland, Liberia, Somalia and Cape Verde.

Calculating the External Terms of Trade for Agriculture

The barter terms of trade were simply calculated as the ratios of a price index of agricultural commodity exports to a price index of agricultural inputs. The indices were constructed using Fischer's ideal index as an aggregate of the exports and imports shown in the table for the respective countries. The data were from the FAO trade statistics from where a price index was derived as a ratio of value to quantity.

Table A2: Agricultural Commodity Exports and Imports Used to Calculate the External Terms of Trade for Agriculture.

Country	Exports	Imports
SSA	Bananas, cocoa, coffee, cotton, groundnuts, rubber, sugar, tea, tobacco.	Wheat, rice, maize, sugar, palm oil, milk.
Burkina Faso	Cotton (69), Cattle.	Wheat, rice.
Cameroon	Cocoa (25), coffee (26), cotton (15), bananas (13), rubber (11).	Wheat, rice, barley.
Côte d'Ivoire	Cocoa (56), coffee (16), cotton (8), rubber (4).	Wheat, rice, cattle, milk.
Mali	Cotton (57), sheep.	Wheat, rice, sugar, milk.
Niger	Cattle.	Wheat, rice, sugar, palm oil, kolanuts.
Ghana	Cocoa (92).	Wheat, rice, sugar.
Nigeria	Cocoa (51), rubber (24).	Wheat, rice, sugar, palm oil, cattle.
Malawi	Tobacco (75), tea (9), coffee (4).	Wheat, maize.
Mozambique	Sugar (27), cotton (26), cashew.	Wheat, maize, sugar, palm oil.
South Africa	Maize, oranges, wool, apples, grapes, sugar, beef.	Wheat, rice.
Tanzania	Coffee (30), cotton (29), tea (9), tobacco (7), cashew.	Wheat, rice, palm oil.
Zimbabwe	Tobacco (60), sugar (6), cotton (5).	Wheat, maize, rice.
Madagascar	Coffee (34), vanilla.	Wheat, maize, palm oil.
Ethiopia	Coffee (67).	Wheat, sorghum, maize.
Kenya	Coffee (22), tea (33).	Wheat, rice, sugar, palm oil.
Uganda	Coffee (77), beans.	Wheat, rice, sugar, palm oil.

Source: FAO Trade Statistics.
(Figures in parenthesis are the percentages of the crop in total agricultural exports. A more detailed composition of exports is provided in Tables A4 and A5).

The net income terms of trade was calculated as

$$net\ income\ terms\ of\ trade\ =\ \frac{P_x}{P_m}Q_x$$

where P_x is the agricultural export price index, P_m is the agricultural import price index (in this case food crop imports) and Q_x is the quantity exported. Tsakok, (1990) provides a more extensive definition and highlights some of the limitations of these measures.

Table A3: Components of the Macroeconomic Policy Stance, 1996-97 (*updated from Adjustment in Africa, 1994*).

Country	Fiscal Policy — Overall fiscal balance Including Grants (% of GDP)	Fiscal Policy (score)	Monetary Policy — Seigniorage %	Seigniorage Score	Inflation %	Inflation Score	Real Interest Rate %	Real Interest Rate Score	Monetary Policy (score)	Exchange Rate Policy — Parallel Market Exchange Rate Premium (%)	(score)	Change in the Real Effective Exchange Rate 1990 to 1996/97 (%)	(score)	Exchange Rate Policy (score)	Overall Macroeconomic Policies (score)
Benin	-0.1	1	0.2	1	6.0	1	1.5	1	1.0			31.3	2	2	1.3
Burkina Faso	-1.9	2	2.5	3	4.2	1	0.4	1	1.7			49.2	1	1	1.6
Burundi	-6.1	3	-0.1	1	28.7	3	-8.0	3	2.3	22.1	2			2	2.4
Cameroon	-1.4	1	-0.9	1	5.9	1	-0.9	1	1.0			77.7	1	1	1.0
Central African Republic	-2.9	2	0.8	2	2.5	1	1.1	1	1.3			58.2	1	1	1.4
Congo, Republic of	-1.1	1	0.7	2	6.8	1	-4.3	2	1.7			28.9	2	2	1.6
Côte d'Ivoire	-2.0	2	1.1	2	4.4	1	1.4	1	1.3			33.0	2	2	1.8
Gabon	4.4	1	-0.2	1	4.1	1	1.0	1	1.0			104.7	1	1	1.0
Gambia, The	-8.2	3	2.1	3	2.5	1	10.2	4	2.7	10.4	2			2	2.6
Ghana	-9.4	4	2.1	3	37.3	3	-7.6	3	3.0	1.6	1			1	2.7
Kenya	-1.6	2	3.9	4	10.1	2	7.9	3	3.0	4.2	1			1	2.0
Madagascar	-3.7	3	1.9	3	12.2	2	-7.9	3	2.7	7.6	1			1	2.2
Malawi	-5.0	3	5.0	4	23.4	2	-8.1	3	3.0	7.8	1			1	2.3
Mali	-1.5	2	1.0	2	2.9	1	0.0	1	1.3			53.1	1	1	1.4
Mauritania	6.3	1	-0.9	1	4.6	1	-	-	1.0	4.9	1			1	1.0
Mozambique	-4.0	3	-	-	25.5	3	0.0	1	2.0	7.8	1			1	2.0
Niger	-1.5	1	0.8	2	5.4	1	1.1	1	1.3			55.9	1	1	1.1
Nigeria	3.3	1	1.0	2	19.9	2	-12.2	4	2.7	273.8	4			4	2.6
Rwanda	-4.2	3	2.3	3	10.5	2	-2.6	1	2.0	8.1	1			1	2.0
Senegal	-0.2	1	0.3	1	2.3	1	3.6	2	1.3			42.7	1	1	1.1
Sierra Leone	-6.9	3	0.6	2	19.0	2	7.1	3	2.3	2.3	1			1	2.3
South Africa	-5.6	3	1.2	2	8.0	1	7.0	3	2.0	3.3	1			1	2.0
Tanzania	-2.1	2	3.9	4	21.4	2	-9.6	3	3.0	4.7	1			1	2.0
Togo	-3.5	3	2.9	3	6.4	1	1.8	1	1.7			34.1	2	2	2.2
Uganda	-1.9	2	0.7	2	7.0	1	2.9	1	1.3	8.5	1			1	1.4
Zambia	-3.9	3	-0.9	1	35.6	3	-0.7	1	1.7	6.5	1			1	1.9
Zimbabwe	-7.6	4	0.8	2	20.4	2	-1.2	1	1.7	7.9	1			1	2.1

Scoring Criteria (as used in *Adjustment in Africa, 1994*)

		Fiscal (% of GDP)	Seigniorage	Inflation	Real Interest Rate	Monetary Policy	Parallel Market Premium	Change REER	Overall Macroeconomic Policies
Good or adequate	1	> -1.5	< 0.5	> 10	< 3	<1.4	< 10	> 40	1 to 1.3
Fair	2	-1.5 to -3.5	0.6 to 1.5	11 to 25	4 to 6	≥1.4<2.4	11 to 30	21 to 40	1.4 to 2.3
Poor	3	-3.6 to -7.0	1.6 to 3	26 to 50	7 to 10	≥2.4<3.0	31 to 50	6 to 20	2.4 to 3.0
Very Poor	4	< -7.1	> 3	> 50	> 10	≥3.0	> 50	< 6	> 3.0

Source: World Bank and IMF data.

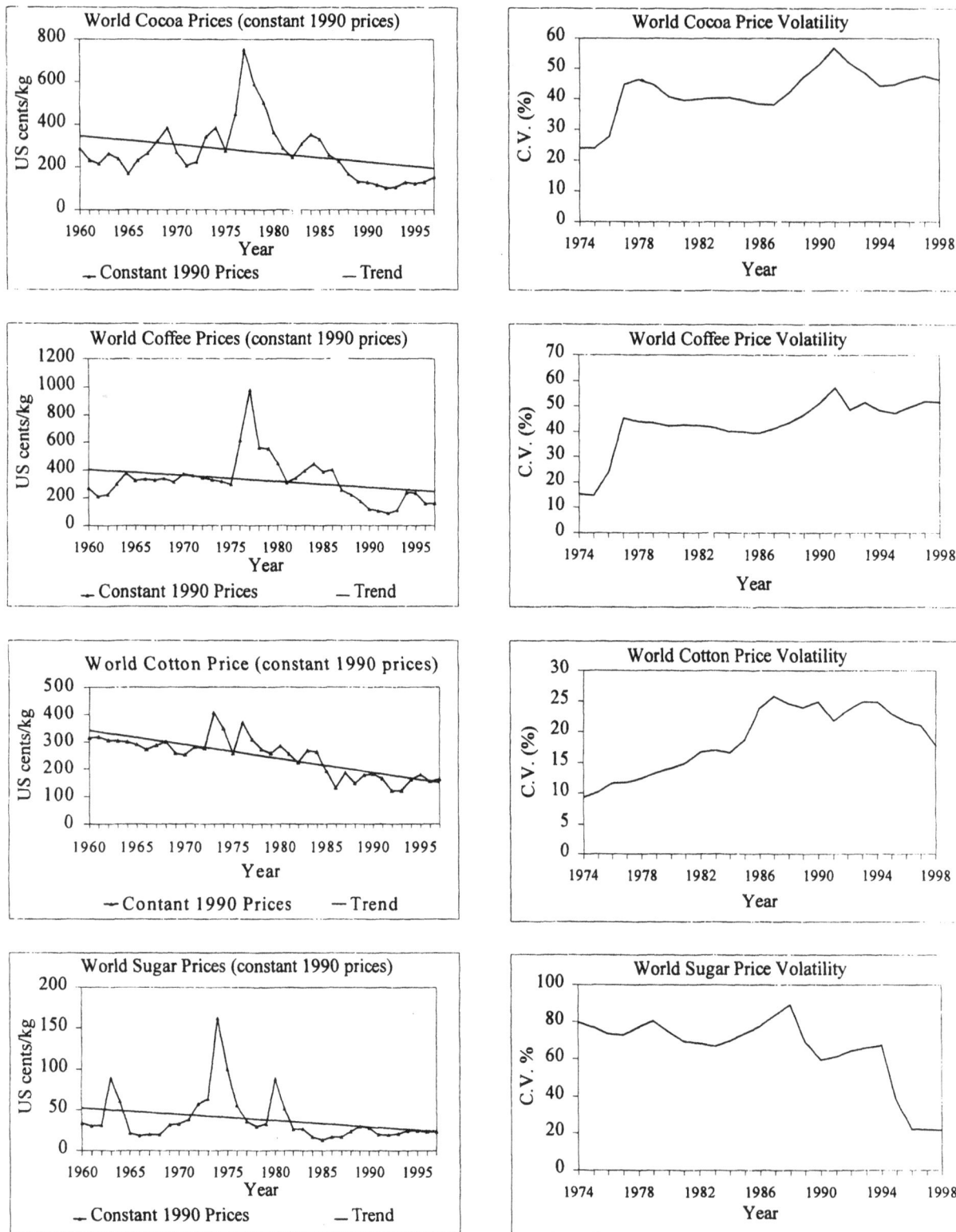

Figure A1: Trends in the Level and Volatility of Real World Agricultural Commodity Prices.

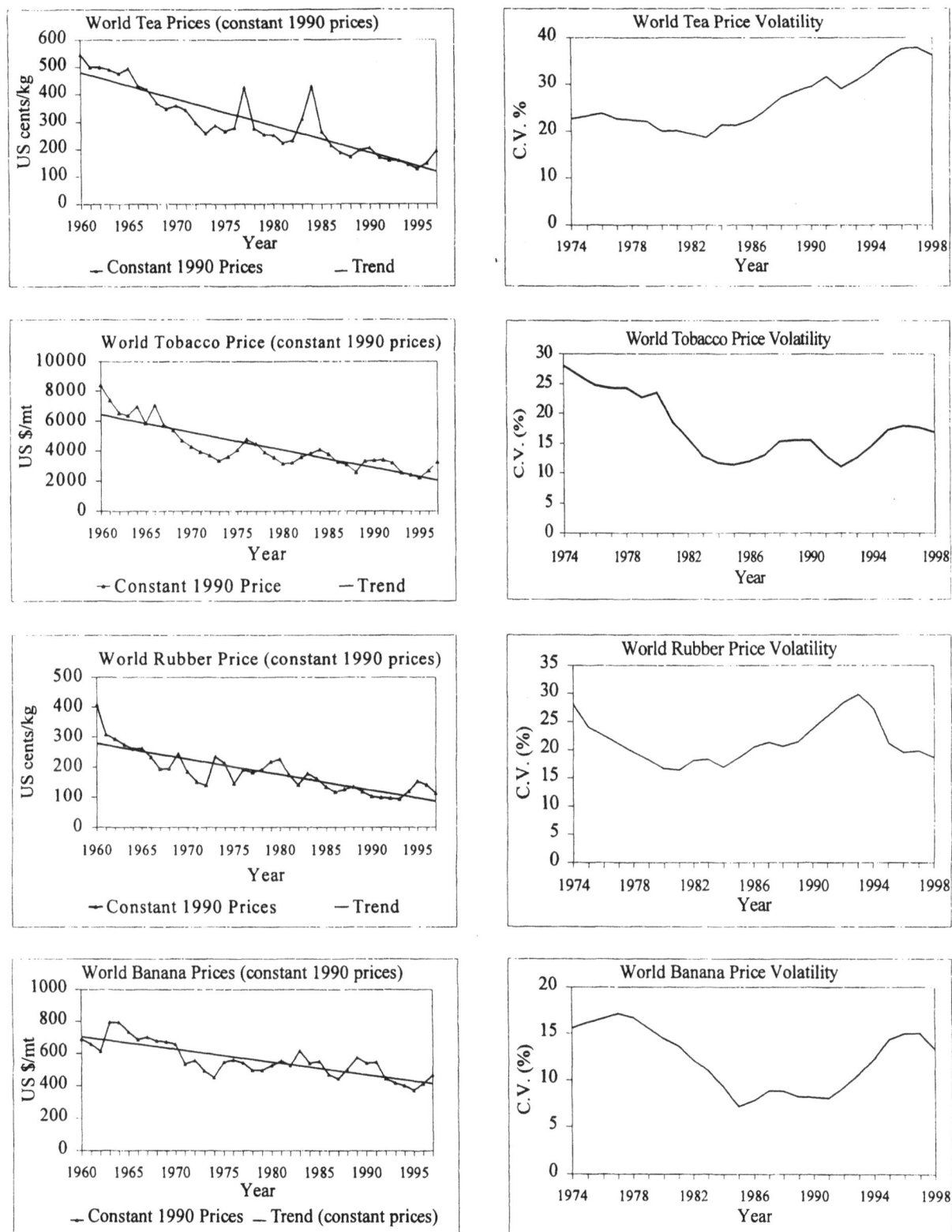

Figure A1: Trends in the Level and Volatility of Real World Agricultural Commodity Prices (*continued*).
The coefficient of variation (C.V.) plots show a 15 year rolling window, e.g. 1974 in the graphs represent the C.V. from 1960 to 1974, 1975 represents the C.V. from 1961 to 1975 etc. World Bank data are used.

Table A4: Sub-Saharan African Country Exports by Commodity, 1993-95.

Country	Bananas	Cocoa	Coffee	Cotton	G'nuts	Rubber	Sugar	Tea	Tobacco	Other
Angola	-	-	71.3	-	-	-	-	-	-	28.7
Benin	-	-	-	93.5	-	-	-	-	-	6.5
Botswana	-	-	-	0.4	-	-	3.1	-	-	96.3
Burkina Faso	-	-	-	69.2	0.2	-	-	-	0.1	30.5
Burundi	-	-	79.3	2.6	-	-	1.2	10.3	1.1	5.4
Cameroon	12.7	25.4	25.6	15.0	-	11.3	0.3	-	0.3	9.4
Cape Verde	82.4	-	0.8	-	-	-	-	-	-	16.7
Central African Republic	-	-	30.7	28.9	0.3	-	0.1	-	1.6	38.5
Chad	-	-	-	63.3	-	-	-	-	-	36.7
Congo, Dem R	-	3.4	72.6	0.7	-	6.0	-	1.3	-	16.0
Congo, Rep	-	1.0	3.3	-	-	0.1	90.8	-	-	4.7
Côte d'Ivoire	3.8	56.3	15.5	7.6	-	4.2	0.8	-	-	11.8
Equatorial Guinea	0.9	97.1	2.0	-	-	-	-	-	-	0.0
Ethiopia	-	-	67.1	1.6	-	-	0.6	0.1	-	30.7
Ethiopia PDR	-	-	66.6	0.2	-	-	1.0	-	-	32.2
Gabon	-	24.2	5.1	-	-	38.5	10.6	-	-	21.6
The Gambia	-	-	-	8.7	51.7	-	0.2	0.1	0.4	39.0
Ghana	0.1	91.7	0.6	1.2	-	1.9	-	-	0.1	4.3
Guinea	-	5.0	29.1	25.3	-	-	-	-	-	40.6
Guinea-Bissau	-	-	-	4.5	0.5	-	-	-	-	95.0
Kenya	-	-	22.4	0.7	-	0.1	2.8	33.2	0.8	40.0
Lesotho	-	-	-	-	-	-	-	-	-	100.0
Liberia	-	1.3	-	0.1	-	91.9	-	-	-	6.7
Madagascar	-	1.8	34.3	0.5	0.1	-	5.6	0.1	0.1	57.4
Malawi	-	-	3.5	0.8	0.1	0.6	7.4	9.1	75.2	3.3
Mali	-	-	-	57.0	0.9	-	-	-	-	42.1
Mauritius	-	-	-	0.1	-	-	89.0	1.3	-	9.6
Mozambique	-	-	-	26.3	0.3	-	28.6	0.3	0.1	44.3
Namibia	-	-	-	-	-	-	-	-	-	100.0
Niger	-	-	-	1.9	-	-	-	-	-	98.1
Nigeria	-	51.0	0.3	2.3	-	24.4	-	-	0.3	21.7
Reunion	-	-	-	-	-	-	89.4	-	-	10.5
Rwanda	-	-	56.4	-	-	-	-	34.2	-	9.4
Sao Tome PDR	-	98.2	0.3	-	-	-	-	-	-	1.5
Senegal	-	-	-	22.9	3.4	-	0.2	-	0.3	73.3
Sierra Leone	-	26.9	33.8	-	-	-	0.7	-	5.4	33.2
Somalia	6.4	-	-	-	-	-	-	-	-	93.6
South Africa	-	-	0.3	0.1	1.0	-	4.8	0.2	1.0	92.7
Sudan	-	-	-	26.4	0.9	-	8.0	-	-	64.8
Swaziland	-	-	-	2.2	-	-	47.5	-	-	50.2
Tanzania	-	0.7	29.9	28.9	0.1	-	1.4	8.8	6.6	23.5
Togo	-	6.5	15.8	59.8	-	-	0.1	-	-	17.8
Uganda	0.1	0.2	77.0	2.2	0.1	-	-	3.3	2.2	14.9
Zambia	-	-	7.5	4.7	0.4	-	47.2	0.1	16.2	23.8
Zimbabwe	-	-	1.7	5.1	0.3	-	6.0	1.2	59.5	26.2

Source: FAO Trade Statistics.

178

Table A5: Sub-Saharan African Commodity Exports by Country, 1993-95.

Country	Bananas	Cocoa	Coffee	Cotton	G'nts	Rubber	Sugar	Tea	Tobacco	Other
Angola	-	-	0.2	-	-	-	-	-	-	-
Benin	-	-	-	11.3	-	-	-	-	-	0.2
Botswana	-	-	-	-	-	-	0.4	-	-	2.3
Burkina Faso	-	-	-	6.8	0.5	-	-	-	-	0.7
Burundi	-	-	5.0	0.2	-	-	0.1	2.1	0.1	0.1
Cameroon	41.3	6.3	7.0	5.8	0.2	19.3	0.1	-	0.2	0.8
Cape Verde	0.5	-	-	-	-	-	-	-	-	-
Central African Republic	-	-	0.8	1.1	0.3	-	-	-	0.1	0.3
Chad	-	-	-	7.4	-	-	-	-	-	1.0
Congo, Dem R	-	0.3	4.7	0.1	-	2.4	-	0.3	-	0.3
Congo, Rep	-	-	-	-	-	-	1.1	-	-	-
Côte d'Ivoire	53.6	61.7	18.7	12.8	0.1	31.7	1.5	-	-	4.6
Equatorial Guinea	-	0.3	-	-	-	-	-	-	-	-
Eritrea	-	-	-	-	-	-	-	-	-	0.1
Ethiopia	-	-	11.4	0.4	-	-	0.2	-	-	1.7
Ethiopia PDR	-	-	2.1	-	-	-	0.1	-	-	0.3
Gabon	-	0.1	-	-	-	1.0	0.1	-	-	-
The Gambia	-	-	-	0.1	17.6	-	-	-	-	0.1
Ghana	0.4	20.1	0.2	0.4	-	2.9	-	-	0.1	0.3
Guinea	-	0.1	0.9	1.1	-	-	-	-	-	0.4
Guinea-Bissau	-	-	-	0.1	0.2	-	-	-	-	0.4
Kenya	0.1	-	15.6	0.6	0.7	0.4	3.1	73.9	1.0	9.0
Lesotho	-	-	-	-	-	-	-	-	-	0.3
Liberia	-	-	-	-	-	9.2	-	-	-	0.1
Madagascar	-	0.2	4.3	0.1	0.2	-	1.1	-	-	2.3
Malawi	-	-	0.9	0.3	0.9	1.0	2.9	7.2	32.2	0.3
Mali	-	-	-	14.4	5.7	-	-	-	-	2.4
Mauritania	-	-	-	-	-	-	-	-	-	0.9
Mauritius	-	-	-	-	-	-	38.1	1.1	-	0.8
Mozambique	-	-	-	1.4	0.4	-	1.7	-	-	0.5
Namibia	-	-	-	-	-	-	-	-	-	4.4
Niger	-	-	-	0.1	-	-	-	-	-	1.1
Nigeria	-	9.8	0.1	0.7	0.2	32.1	-	-	0.1	1.5
Reunion	-	-	-	-	-	-	13.5	-	-	0.3
Rwanda	-	-	1.7	-	-	-	-	3.3	-	0.1
Sao Tome PDR	-	0.2	-	-	-	-	-	-	-	-
Senegal	-	-	-	2.6	9.4	-	-	-	-	1.9
Sierra Leone	-	0.2	0.3	-	-	-	-	-	0.1	0.1
Somalia	3.5	-	-	-	-	-	-	-	-	1.4
South Africa	0.1	-	0.4	0.2	46.2	0.1	10.2	0.7	2.4	40.7
Sudan	-	-	-	10.9	9.0	-	3.7	-	-	6.2
Swaziland	-	-	-	0.6	-	-	15.0	-	-	3.3
Tanzania	-	0.2	7.3	9.9	1.1	-	0.2	6.9	2.8	1.9
Togo	-	0.4	1.1	5.9	-	-	-	-	-	0.4
Uganda	0.3	-	16.3	0.7	0.6	-	-	2.2	0.8	1.0
Zambia	-	-	0.1	0.1	0.3	-	1.4	-	0.6	0.1
Zimbabwe	0.1	-	1.0	4.1	6.5	-	5.4	2.2	59.5	4.8
Total	100	100	100	100	100	100	100	100	100	100

Source: FAO Trade Statistics.

179

Table A6: Producer Price Trends for Sub-Saharan Africa's Major Export Crops.

Commodity	Country	%[32]	Annual %[33] Change in Real Producer Price (1990-96/97)	Producer's Share of f.o.b. Price (% 1996-97)	% Change in Producer's Share of f.o.b. Price (1990-96/97)	f.o.b. Price as a Share of world Price (% 1996-97)	% Change in f.o.b Share of World Price (1990-96/97)
Cocoa	World[34]		**2.8**	-	-	-	-
(98%)[35]	Côte d'Ivoire	62	3.2	46	-23	96	2
	Ghana	20	0.7	39	-5	96	-10
	Nigeria	10	6.8	98	9	92	8
	Cameroon	6	2.5	76	8	86	-5
Coffee	World		**9.9**	-	-	-	-
(82%)	Côte d'Ivoire	19	9.7	72	2	92	1
	Uganda	17	8.8	72	29	78	9
	Kenya	16	8.2	81	5	95	0
	Ethiopia	12	-	-	-	-	-
	Cameroon	7	9.8	73	12	87	-1
	Tanzania	7	14.1	77	18	98	1
	Madagascar	4	12.9	70	-9	81	15
Cotton	World		**0.5**	-	-	-	-
(82%)	Mali	15	0.8	44	-3	82	-5
	Côte d'Ivoire	13	0.4	47	-5	88	-5
	Benin	11	2.0	37	-15	95	5
	Tanzania	10	5.9	64	21	83	-10
	Burkina Faso	7	1.5	35	-28	96	20
	Chad	7	-0.6	36	-19	92	2
	Cameroon	6	4.5	51	-2	82	-6
	Togo	6	1.4	39	-12	100	5
	Zimbabwe	4	13.8	88	32	99	1
	Senegal	3	3.2	47	-15	97	-1
Groundnuts	World		**0.8**	-	-	-	-
(81%)	South Africa	46	-0.6	82	12	88	-3
	The Gambia	18	-0.1	65	18	52	-21
	Senegal	10	4.0	62	-1	75	-5
	Zimbabwe	7	-1.0	68	-1	105	11
Sugar	World		**0.8**	-	-	-	-
(82%)	Mauritius	38	-1.5	94	13	162	19
	Swaziland	15	-1.8	65	-15	143	24
	South Africa	10	4.5	92	9	75	-3
Tea	World		**-2.0**	-	-	-	-
(88%)	Kenya	74	-6.3	89	1	77	-8
	Malawi	7	-0.8	92	8	84	-7
	Tanzania	7	-7.3	40	0	66	7
Tobacco	World		**-3.1**	-	-	-	-
(95%)	Zimbabwe	60	-2.5	79	1	93	-1
	Malawi	32	4.2	82	7	86	-2
	Tanzania	3	·9.3	57	19	61	1

Source: World Bank and IMF data.

---

[32] This is the percentage of the individual countries exports in total Sub-Saharan Africa exports.

[33] These are usually referred to as growth rates but in the case of prices they will be referred to as annual percentage changes.

[34] World prices are in $US, other prices are in local currency.

[35] This represents the percentage of total Sub-Saharan exports of this commodity that is accounted for by countries listed. In the case of cocoa, Côte d'Ivoire, Ghana, Nigeria and Cameroon account for 98 percent of all cocoa exported from Sub-Saharan Africa.

Table A7: Producer Price Trends for Export Crops by Country in Sub-Saharan Africa.

Country	Commodity	Annual % Change in Real Producer Price (1990-96/97)	Producer's Share of f.o.b. Price % (1996-97)	% Change in Producer's Share of f.o.b. Price (1990-96/97)	f.o.b. Price as a Share of world Price % (1996-97)	% Change in f.o.b Share of World Price (1990-96/97)
Benin	Cotton	2.0	37	-15	95	5
Botswana	Cattle	-7.6	-	-	-	-
Burkina Faso	Cotton	1.5	35	-28	96	20
Burundi	Coffee	-2.8	32	-27	98	9
Cameroon	Cocoa	5.6	76	8	86	-5
	Coffee	9.8	73	12	87	-1
	Cotton	4.5	51	-2	82	-6
Chad	Cotton	-0.6	36	-19	92	2
Côte d'Ivoire	Cocoa	3.2	46	-23	96	2
	Coffee	9.7	72	2	92	1
	Cotton	0.4	47	-5	88	-5
Ethiopia	Coffee	-0.1	-	-	-	-
The Gambia	Groundnuts	-0.1	65	18	52	-21
Ghana	Cocoa	0.7	39	-5	96	-10
Kenya	Coffee	8.2	81	5	95	0
	Tea	-6.3	89	1	77	-8
Madagascar	Coffee	12.9	70	-9	81	15
	Vanilla	2.1	62	7	114	-7
Malawi	Tobacco	4.2	82	7	86	-2
	Tea	-0.8	92	8	84	-7
Mali	Cotton	0.8	44	-3	82	-5
Mauritius	Sugar	-1.5	94	13	162	19
Mozambique	Cotton	3.8	64	6	73	-16
	Cashew	5.9	55	1	85	9
Niger	Cattle	22.3	-	-	-	-
Nigeria	Cocoa	6.8	98	9	92	8
	Rubber	15.2	100	16	93	2
Senegal	Cotton	3.2	47	-15	97	-1
	Groundnuts	4.0	62	-1	75	-5
South Africa	Maize	-1.5	93	-6	86	-7
	Oranges	-4.9	70	-12	75	-5
	Apples	-0.2	93	16	87	-1
	Sugar	4.5	92	9	75	-3
	Wool	-5.8	89	-4	101	23
	Beef	-2.1	75	5	110	-9
Swaziland	Sugar	-1.8	65	-15	143	24
Tanzania	Coffee	14.1	77	18	98	1
	Cotton	5.9	64	21	83	-10
	Tea	-7.3	40	0	66	7
	Tobacco	9.3	57	19	61	1
	Cashew	1.7	71	19	102	-1
Togo	Cotton	1.4	39	-12	100	5
Uganda	Coffee	8.8	72	29	78	9
Zimbabwe	Tobacco	-2.5	79	1	93	-1
	Cotton	13.8	88	32	99	1

Source: World Bank and IMF data.

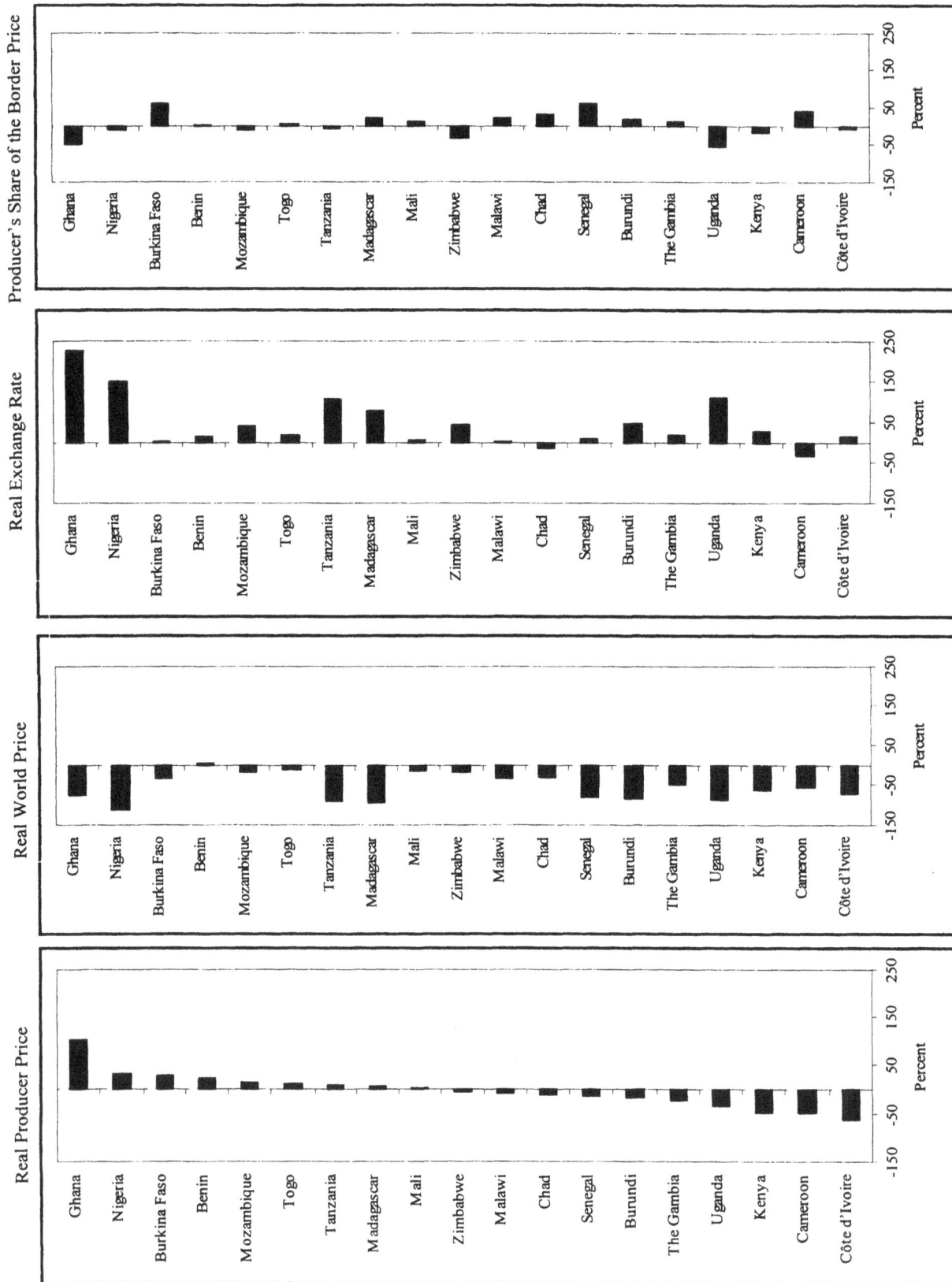

Figure A2: The Structure of Agricultural Export Price Incentives, 1981-89 to 1989-91.
Source: Table A8.

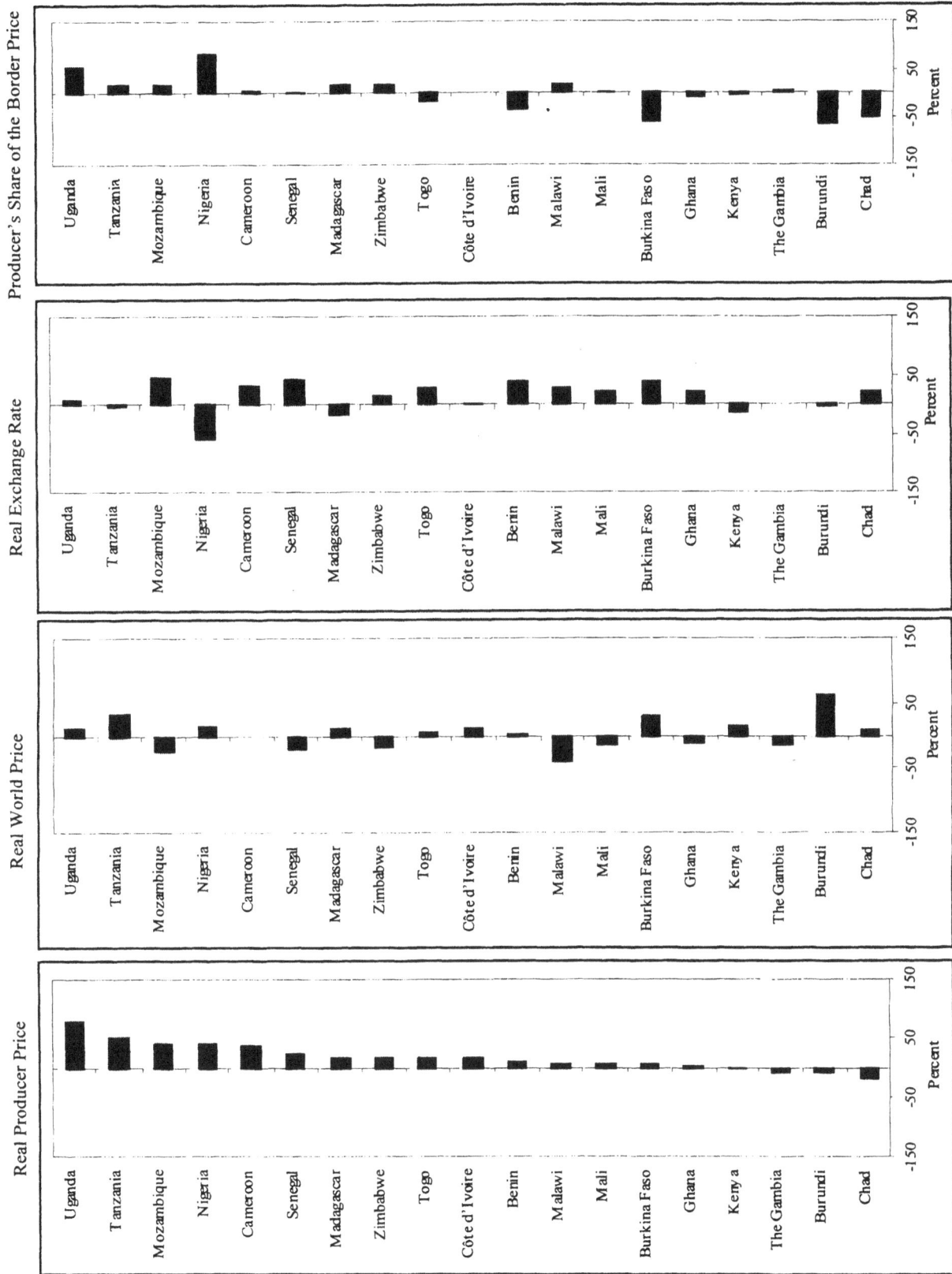

Figure A3: The Structure of Agricultural Export Price Incentives, 1989-91 to 1995-97.
Source: Table A8.

183

Table A8: Real Producer Price Decomposition.

Country	Period	Real Producer Price (% Change)	External Factors Real World Price (% Change)	Domestic Policy Factors Aggregate Effect (% Changes)	RER (% Change)	NPC (% Change)
Benin	1	22	5	17	15	2
	2	8	2	6	40	-34
	3	30	7	23	55	-32
Burkina Faso	1	28	-34	62	3	59
	2	6	29	-23	39	-62
	3	34	-5	39	42	-3
Burundi	1	-18	-83	64	46	18
	2	-10	60	-70	-5	-65
	3	-28	-22	-5	42	-47
Cameroon	1	-48	-56	8	-33	41
	2	36	0	36	33	3
	3	-12	-56	44	0	44
Côte d'Ivoire	1	-63	-71	8	16	-8
	2	15	13	2	2	0
	3	-47	-58	11	18	-7
Chad	1	-12	-29	17	-15	32
	2	-18	9	-27	23	-50
	3	-30	-20	-10	8	-18
The Gambia	1	-21	-49	28	17	11
	2	-9	-13	4	1	3
	3	-30	-61	32	18	14
Ghana	1	102	-75	177	227	-50
	2	3	-11	13	21	-8
	3	105	-86	191	249	-58
Kenya	1	-47	-60	13	29	-16
	2	-3	14	-17	-13	-4
	3	-50	-46	-4	16	-20
Madagascar	1	6	-93	100	77	23
	2	16	12	4	-16	20
	3	22	-81	103	60	43
Malawi	1	-9	-33	24	2	22
	2	7	-39	46	31	15
	3	-2	-72	71	33	38
Mali	1	4	-12	16	4	12
	2	7	-12	19	21	-2
	3	11	-24	35	25	10
Mozambique	1	14	-16	29	39	-10
	2	41	-22	63	45	18
	3	54	-38	92	84	8

Table A8: Real Producer Price Decomposition *(continued).*

Country	Period	Real Producer Price (% Change)	External Factors	Domestic Policy Factors		
			Real World Price (% Change)	Aggregate effect (% Changes)	RER (% Change)	NPC (% Change)
Nigeria	1	31	-111	142	151	-9
	2	41	14	27	-58	85
	3	72	-97	169	93	76
Senegal	1	-13	-81	68	9	59
	2	22	-21	43	44	-1
	3	9	-102	111	53	58
Tanzania	1	9	-90	99	106	-7
	2	49	34	15	-5	20
	3	58	-56	114	101	13
Togo	1	13	-9	22	18	4
	2	15	6	9	28	-19
	3	29	-3	32	46	-14
Uganda	1	-33	-88	55	109	-54
	2	77	11	67	9	58
	3	45	-77	123	118	5
Zimbabwe	1	-6	-18	12	44	-32
	2	16	-16	33	13	20
	3	10	-34	44	57	-13

Period 1 is 1981-83 to 1989-91; Period 2 is 1989-91 to 1995-97 and Period 3 is 1981-83 to 1995-97.
Source: Calculated from World Bank, IMF data and FAO data.

Table A9: Fertilizer Policy Scores.

Country	Import Tariff (%)	Score	Non-tariff Barriers	Score	Pricing Arrangements	Score	Av. Score	Score Conversion (0 to 100)
Benin	6.00	4	Government controls the type of fertilizer that can be imported, specifying a list of fertilizers that can be sold. Government also controls who imports the fertilizer and specifies the distribution of each companies fertilizer quota to specific regions.	2	The government continues to fix the price of fertilizer pan-territorially based on a formula using the c.i.f. international price plus a marketing and profit margin.	3	3.00	60
Burkina Faso	10.31	3	The dualistic nature of the fertilizer market restricts private traders to the non-cotton areas. SOFITEX procures fertilizer on international markets and provides fertilizer to cotton farmers, SOFITEX is responsible for the whole distribution system. DIMA distributes aid-funded fertilizer to non-cotton regions using private traders.	4	Prices of fertilizer aid have been set below market levels, particularly after 1994 devaluation. Most of this fertilizer was sold domestically and traded to neighboring countries.	4	3.67	73
The Gambia	Low	5	Uncertainty of the future role of governmental and international donors has inhibited the response by private entrepreneurs in the fertilizer market.	3	Fertilizer prices at the wholesale and retail levels are based on world market prices.	5	4.33	87
Cameroon	5.85	4	-	3	Fertilizer prices are market determined.	5	4.00	80
Côte d'Ivoire	0.37	5	-	3	Fertilizer prices are market determined.	5	4.33	87
Ghana	0.00	5	Government controls the types of fertilizer the can be imported, specifying a list of fertilizers that can be sold. Private traders need prior government approval before importing fertilizers.	2	Fertilizer prices are market determined.	5	4.00	80
Kenya	Low	5	Licensing of importers and distributors have been removed. Large volumes of fertilizer aid distort market prices. Government owned distributors continue to dominate the market.	3	Fertilizer prices are market determined.	5	4.33	87
Malawi	0.04	5	The fertilizer market (both imports and distribution) has been liberalized. The Small-holder Farmer Fertilizer Revolving Fund of Malawi (SFFRFM) remains the main importer of fertilizer.	4	Fertilizer prices are market determined. There is some concern that fertilizer subsidies will be re-introduced.	5	4.67	93

Table A9: Fertilizer Policy Scores *(continued)*.

Country	Import Tariff (%)	Score	Non-tariff Barriers	Score	Pricing Arrangements	Score	Av. Score	Score Conversion (0 to 100)
Mali	5.00	4	The same fertilizers have been continuously imported for a long period of time and the input delivery system is largely government controlled or semi-government controlled (cotton organizations).	3	Fertilizer prices are market determined.	5	4.00	80
Nigeria	Low	4	Fertilizer is sold to neighboring countries due to continuing fertilizer subsidies.	4	Fertilizer subsidies are still prevalent.	3	3.67	73
Senegal	20.82	1	SENCHIM, the main commercial distributing branch of Industrie Chimique du Senegal (ICS), dominates the fertilizer market.	4	Fertilizer prices are market determined.	5	3.33	67
South Africa	Low	5	The fertilizer market is concentrated with a few large producers and distributors.	4	Fertilizer prices are market determined.	5	4.67	93
Tanzania	5.00	4	Absence of control on the quality of fertilizer. Deterioration and fake packaging frequently occurs.	3	Fertilizer prices are market determined.	5	4.00	80
Togo	14.00	3	SOTOCO dominates the fertilizer market for both cotton and maize. Food crop fertilizers are available from SOTOCO or through DRDR agents, however, they can only be bought on credit by cotton growers.	3	Fertilizer prices are largely market determined. Since 1995 large private purchases have been sold at unsubsidized prices.	4	3.33	67
Uganda	0.00	5	The fertilizer market has been liberalized.	4	Fertilizer prices are market determined. Subsidies on donor fertilizers distorts the market.	4	4.33	87
Zimbabwe	3.50	5	Government continues to manage fertilizer imports through permits. A mandatory government approval of imported fertilizer compositions is required.	3	Fertilizer prices are market determined.	5	4.33	87

Tariff: A score of 1 reflects an import tariff rate greater than 20%; 2, 15%-20%; 3, 10%-15%; 4, 5%-10%; 5, 0%-5%; Non-tariff barriers: qualitative evaluation, 1 excessive non-tariff barriers...5 no barriers, Pricing arrangements: qualitative evaluation, 1 excessive intervention...5 no intervention.
Source: World Bank Country Reports and World Bank Staff.

187

Table A10: Price Elasticities of Supply for Selected Export Crops in Sub-Saharan Africa.

Crop	Year	Author	Price Elasticity Short Run	Long Run
Coffee			*0.64*	*1.48*
Kenya (smallholders)	1947-64	Behrman	0.64	1.48
Cotton			*0.26-0.83*	*0.38-0.86*
11 countries in SSA	1970-87	Jaeger	0.67	-
Nigeria	1950-64	Diejomaoh	0.67	0..67
Sudan	1951-65	Medani	0.39	0.50
Uganda	1945-66	Alibaruho	0.50	0.63
Zimbabwe (smallholders)	1975-89	Townsend *et al*	0.83	0.86
Nigeria	1970-86	Kwanashiwe *et al*	1.35	-
Tanzania	1965-84	Mshomba	0.26	0.38
Ghana	1968-81	Senei	0.55	1.32
Cocoa			*0.04-0.68*	*0.71-1.81*
14 countries in SSA	1970-87	Jaeger	0.23	-
Ghana	1947-64	Behrman	0.39	0.77
Nigeria	1947-64	Behrman	-	0.71
Côte d'Ivoire	1947-64	Behrman	-	0.81
Cameroon	1947-64	Behrman	0.68	1.81
Nigeria	1970-86	Kwanashiwe *et al*	0.04	-
Groundnuts			*0.24-0.79*	*0.24-0.79*
Nigeria	1948-67	Olayide	0.24-0.79	0.24-0.79
Rubber			*0.04*	*1.75*
Nigeria	1952-72	Olayemi *et al*	0.04	1.75
Nigeria	1970-86	Kwanashiwe *et al*	0.07	
Tea				
4 countries in SSA	1970-87	Jaeger	-0.04	-
Tanzania	1964-84	Mshomba	0.35	-
Tobacco			*0.28-0.60*	*0.82-1.36*
Nigeria	1945-64	Adesimi	0.60	0.82
Malawi	1946-64	Dean	0.48	0.48
Zimbabwe	1970-89	Townsend *et al*	0.28	1.36
Malawi	1964-89	Chembezi	0.33	0.95

Source: Adapted from Mamingi (1996), Townsend and Thirtle (1997) and Oyejide (1990).

Table A11: Production Distribution of Food Crops Across Sub-Saharan African Countries, 1995-1997 (percent).

Country	Yams	Sorghum	Maize	Wheat	Cassava	Millet	Rice
Angola	-	-	1.0	0.1	2.8	-	0.2
Benin	4.3	0.6	1.5	-	1.6	0.2	0.2
Botswana	-	0.2	-	-	-	-	-
Burkina Faso	0.1	6.7	0.7	-	-	6.3	0.9
Burundi	-	0.4	0.5	0.2	0.6	0.1	0.3
Cameroon	0.4	2.4	2.0	-	1.8	0.6	0.5
Cape Verde	-	-	-	-	-	-	-
Central African Republic	1.1	0.1	0.2	-	0.6	0.1	0.1
Chad	0.8	2.4	0.2	-	0.3	2.0	0.8
Comoros	-	-	-	-	0.1	-	0.2
Congo, DEM R	1.0	0.3	3.3	0.2	20.9	0.3	4.0
Côte d'Ivoire	8.4	0.1	1.7	-	2.0	0.5	8.7
Djibouti	-	-	-	-	-	-	-
Equatorial Guinea	-	-	-	-	0.1 ·	-	-
Eritrea	-	0.4	-	0.2	-	0.2	-
Ethiopia	0.9	10.0	8.8	33.6	-	2.6	-
Gabon	0.5	-	0.1	-	0.3	-	-
The Gambia	-	0.1	-	-	-	0.5	0.2
Ghana	7.3	1.9	3.0	-	8.1	1.6	2.0
Guinea	0.3	-	0.2	-	0.8	0.1	6.2
Guinea-Bissau	-	0.1	-	-	-	0.2	1.2
Kenya	-	0.6	7.6	6.1	1.0	0.4	0.5
Lesotho	-	0.1	0.3	0.3	-	-	-
Liberia	0.1	-	-	-	0.3	-	0.8
Madagascar	-	-	0.5	0.1	2.8	-	23.5
Malawi	-	0.2	4.6	-	0.2	0.1	0.6
Mali	-	3.7	0.8	0.1	-	6.2	4.7
Mauritania	-	0.6	-	-	-	-	0.6
Mauritius	-	-	-	-	-	-	-
Mozambique	-	1.3	2.7	-	5.6	0.3	1.4
Namibia	-	-	0.1	0.1	-	0.5	-
Niger	-	2.0	-	0.1	0.3	14.3	0.6
Nigeria	72.3	36.8	15.5	0.9	36.8	44.5	29.1
Rwanda	-	0.5	0.2	0.1	0.3	-	-
Sao Tome	-	-	-	-	-	-	-
Senegal	-	0.7	0.3	-	0.1	5.2	1.4
Seychelles	-	-	-	-	-	-	-
Sierra Leone	-	0.1	-	-	0.3	0.2	3.6
Somalia	-	0.8	0.4	-	0.1	-	-
South Africa	-	2.1	23.5	41.9	-	0.1	-
Sudan	0.4	17.9	0.2	9.7	-	3.9	-
Swaziland	-	-	0.3	-	-	-	-
Tanzania	-	3.5	7.3	1.1	7.3	2.9	6.2
Togo	1.9	0.9	1.1	-	0.7	0.5	0.5
Uganda	-	1.8	2.4	0.2	2.7	4.2	0.7
Zambia	-	0.2	3.1	1.0	0.6	0.5	0.1
Zimbabwe	-	0.4	5.6	4.0	0.2	0.7	-
Total	100	100	100	100	100	100	100

Source: FAO Production Statistics.

Table A12: Production Distribution of Food Crops within Sub-Saharan African Countries, 1995-1997 (percent).

Country	Yams	Sorghum	Maize	Wheat	Cassava	Millet	Rice	Total
Angola	-	-	11.9	0.2	87.0	-	0.9	100
Benin	39.0	3.2	15.5	-	40.8	0.8	0.6	100
Botswana	-	71.1	21.9	1.7	-	5.3	-	100
Burkina Faso	1.7	51.9	10.0	-	0.1	32.3	4.1	100
Burundi	1.0	8.7	18.5	1.1	64.8	1.5	4.3	100
Cameroon	4.3	15.5	23.0	0.0	52.9	2.4	2.0	100
Cape Verde	-	-	76.0	-	24.0	-	-	100
Central African Republic	33.7	2.5	7.7	-	53.7	1.1	1.4	100
Chad	18.1	33.1	4.7	0.1	19.2	18.5	6.3	100
Comoros	-	-	5.4	-	70.4	-	24.2	100
Congo, DEM R	1.5	0.3	5.7	0.0	90.1	0.2	2.2	100
Côte d'Ivoire	43.9	0.4	9.8	-	28.8	1.0	16.1	100
Djibouti	-	-	100.0	-	-	-	-	100
Equatorial Guinea	-	-	-	-	100.0	-	-	100
Eritrea	-	58.0	6.6	9.0	-	26.4	-	100
Ethiopia	3.6	25.5	40.7	25.6	-	4.5	-	100
Gabon	35.9	-	7.9	-	56.0	-	0.2	100
The Gambia	-	12.2	10.5	-	5.5	54.9	16.8	100
Ghana	20.5	3.3	9.4	-	63.1	1.8	2.0	100
Guinea	6.2	0.3	5.3	-	43.3	0.5	44.4	100
Guinea-Bissau	-	9.5	6.4	-	5.8	13.5	64.8	100
Kenya	-	3.0	64.4	8.5	21.5	1.3	1.4	100
Lesotho	-	12.6	77.2	10.2	-	-	-	100
Liberia	6.2	-	-	-	68.6	-	25.2	100
Madagascar	-	0.0	3.5	0.2	47.1	-	49.3	100
Malawi	-	2.5	82.8	0.1	10.3	1.0	3.4	100
Mali	0.5	30.5	12.2	0.1	0.1	34.3	22.3	100
Mauritania	1.3	59.9	3.7	0.2	0.1	3.1	31.8	100
Mauritius	-	-	98.8	-	-	-	1.2	100
Mozambique	-	4.1	14.9	0.0	77.9	0.7	2.4	100
Namibia	-	6.8	25.5	4.0	-	63.7	-	100
Niger	-	15.3	0.2	0.2	9.2	72.2	2.8	100
Nigeria	29.7	9.3	7.1	0.1	42.2	7.5	4.2	100
Rwanda	1.0	25.2	17.7	1.5	53.4	0.2	1.0	100
Sao Tome	16.7	-	47.4	-	35.8	-	-	100
Senegal	-	12.3	9.3	-	4.0	60.1	14.3	100
Seychelles	-	-	-	-	100.0	-	-	100
Sierra Leone	-	3.0	1.2	-	37.5	3.1	55.1	100
Somalia	-	43.2	41.6	0.3	14.3	-	0.6	100
South Africa	-	3.7	74.2	21.9	-	0.1	0.0	100
Sudan	2.8	73.2	1.2	11.8	0.2	10.7	0.0	100
Swaziland	-	1.4	97.3	0.3	-	-	0.9	100
Tanzania	0.1	6.3	23.7	0.6	59.4	3.5	6.4	100
Togo	32.3	9.0	20.7	-	31.5	3.4	3.1	100
Uganda	-	8.3	20.1	0.2	56.3	13.1	2.0	100
Zambia	-	1.8	60.1	3.3	30.7	3.3	0.7	100
Zimbabwe	-	3.3	77.7	9.1	6.3	3.5	0.0	100

Source: FAO Production Statistics.

Table A13: Fertilizer Consumption (100 grams) per Hectare of Arable Land.

Country Name	1980	1985	1990	1994	% of Other Developing Country Average	% Change 1990-94
China	1530	1724	2798	3087		10
India	855	611	602	933		55
Indonesia	328	503	742	796		7
Brazil	451	717	746	847		14
Simple Average	**545**	**610**	**697**	**859**		**23**
Mauritius	2492	2615	2616	2754	320.7	5
Swaziland	1075	536	661	696	81.1	5
South Africa	803	667	601	631	73.5	5
Zimbabwe	676	605	622	593	69.1	-5
Kenya	144	243	258	305	35.5	18
Malawi	250	231	287	214	24.9	-25
Mauritania	67	100	190	192	22.4	1
Lesotho	154	115	145	188	21.9	30
Côte d'Ivoire	172	116	97	170	19.8	75
Nigeria	57	94	125	120	14.0	-4
Tanzania	125	134	146	114	13.3	-22
Zambia	154	155	113	112	13.0	-1
Congo	35	290	94	112	13.0	19
Benin	5	63	59	91	10.6	54
Senegal	83	87	51	85	9.9	67
Mali	69	95	73	84	9.8	15
Burkina Faso	15	40	59	65	7.6	10
Sudan	65	74	69	56	6.5	-19
Sierra Leone	36	68	24	56	6.5	133
Gambia, The	127	231	32	47	5.5	47
Togo	11	42	48	46	5.4	-4
Cameroon	46	81	24	43	5.0	79
Madagascar	29	32	35	36	4.2	3
Burundi	9	20	18	26	3.0	44
Botswana	35	11	21	24	2.8	14
Ghana	34	31	33	23	2.7	-30
Mozambique	90	12	8	22	2.6	175
Chad	3	23	18	21	2.4	17
Guinea-Bissau	7		17	18	2.1	6
Guinea	4	5	16	15	1.7	-6
Gabon	2	62	25	9	1.0	-64
Rwanda	1	13	26	9	1.0	-65
Central African Republic	7	15	4	6	0.7	50
Zaire	10	9	8	5	0.6	-38
Uganda	1	0	0	4	0.5	-
Niger	8	10	6	3	0.3	-50
Simple Average	**192**	**198**	**184**	**194**	**23**	**15**

Sources: World Development Indicators, 1997.

Distributors of World Bank Group Publications

Prices and credit terms vary from country to country. Consult your local distributor before placing an order.

ARGENTINA
World Publications SA
Av. Cordoba 1877
1120 Ciudad de Buenos Aires
Tel: (54 11) 4815-8156
Fax: (54 11) 4815-8156
E-mail: wpbooks@infovia.com.ar

AUSTRALIA, FIJI, PAPUA NEW GUINEA, SOLOMON ISLANDS, VANUATU, AND SAMOA
D.A. Information Services
648 Whitehorse Road
Mitcham 3132, Victoria
Tel: (61) 3 9210 7777
Fax: (61) 3 9210 7788
E-mail: service@dadirect.com.au
URL: http://www.dadirect.com.au

AUSTRIA
Gerold and Co.
Weihburggasse 26
A-1011 Wien
Tel: (43 1) 512-47-31-0
Fax: (43 1) 512-47-31-29
URL: http://www.gerold.co/at.online

BANGLADESH
Micro Industries Development
Assistance Society (MIDAS)
House 5, Road 16
Dhanmondi R/Area
Dhaka 1209
Tel: (880 2) 326427
Fax: (880 2) 811188

BELGIUM
Jean De Lannoy
Av. du Roi 202
1060 Brussels
Tel: (32 2) 538-5169
Fax: (32 2) 538-0841

BRAZIL
Publicações Tecnicas Internacionais Ltda.
Rua Peixoto Gomide, 209
01409 Sao Paulo, SP.
Tel: (55 11) 259-6644
Fax: (55 11) 258-6990
E-mail: postmaster@pti.uol.br
URL: http://www.uol.br

CANADA
Renouf Publishing Co. Ltd.
5369 Canotek Road
Ottawa, Ontario K1J 9J3
Tel: (613) 745-2665
Fax: (613) 745-7660
E-mail:
order.dept@renoufbooks.com
URL: http://www.renoufbooks.com

CHINA
China Financial & Economic
Publishing House
8, Da Fo Si Dong Jie
Beijing
Tel: (86 10) 6401-7365
Fax: (86 10) 6401-7365

China Book Import Centre
P.O. Box 2825
Beijing

Chinese Corporation for Promotion
of Humanities
52, You Fang Hu Tong,
Xuan Nei Da Jie
Beijing
Tel: (86 10) 660 72 494
Fax: (86 10) 660 72 494

COLOMBIA
Infoenlace Ltda.
Carrera 6 No. 51-21
Apartado Aereo 34270
Santafé de Bogotá, D.C.
Tel: (57 1) 285-2798
Fax: (57 1) 285-2798

COTE D'IVOIRE
Center d'Edition et de Diffusion
Africaines (CEDA)
04 B.P. 541
Abidjan 04
Tel: (225) 24 6510; 24 6511
Fax: (225) 25 0567

CYPRUS
Center for Applied Research
Cyprus College
6, Diogenes Street, Engomi
P.O. Box 2006
Nicosia
Tel: (357 2) 59-0730
Fax: (357 2) 66-2051

CZECH REPUBLIC
USIS, NIS Prodejna
Havelkova 22
130 00 Prague 3
Tel: (420 2) 2423 1486
Fax: (420 2) 2423 1114
URL: http://www.nis.cz/

DENMARK
SamfundsLitteratur
Rosenoerns Allé 11
DK-1970 Frederiksberg C
Tel: (45 35) 351942
Fax: (45 35) 357822
URL: http://www.sl.cbs.dk

ECUADOR
Libri Mundi
Libreria Internacional
P.O. Box 17-01-3029
Juan Leon Mera 851
Quito
Tel: (593 2) 521-606; (593 2) 544-185
Fax: (593 2) 504-209
E-mail: librimu1@librimundi.com.ec
E-mail: librimu2@librimundi.com.ec

CODEU
Ruiz de Castilla 763, Edif. Expocolor
Primer piso, Of. #2
Quito
Tel/Fax: (593 2) 507-383; 253-091
E-mail: codeu@impsat.net.ec

EGYPT, ARAB REPUBLIC OF
Al Ahram Distribution Agency
Al Galaa Street
Cairo
Tel: (20 2) 578-6083
Fax: (20 2) 578-6833

The Middle East Observer
41, Sherif Street
Cairo
Tel: (20 2) 393-9732
Fax: (20 2) 393-9732

FINLAND
Akateeminen Kirjakauppa
P.O. Box 128
FIN-00101 Helsinki
Tel: (358 0) 121 4418
Fax: (358 0) 121-4435
E-mail: akatilaus@stockmann.fi
URL: http://www.akateeminen.com

FRANCE
Editions Eska; DBJ
48, rue Gay Lussac
75005 Paris
Tel: (33-1) 55-42-73-08
Fax: (33-1) 43-29-91-67

GERMANY
UNO-Verlag
Poppelsdorfer Allee 55
53115 Bonn
Tel: (49 228) 949020
Fax: (49 228) 217492
URL: http://www.uno-verlag.de
E-mail: unoverlag@aol.com

GHANA
Epp Books Services
P.O. Box 44
TUC
Accra
Tel: 223 21 778843
Fax: 223 21 779099

GREECE
Papasotiriou S.A.
35, Stournara Str.
106 82 Athens
Tel: (30 1) 364-1826
Fax: (30 1) 364-8254

HAITI
Culture Diffusion
5, Rue Capois
C.P. 257
Port-au-Prince
Tel: (509) 23 9260
Fax: (509) 23 4858

HONG KONG, CHINA; MACAO
Asia 2000 Ltd.
Sales & Circulation Department
302 Seabird House
22-28 Wyndham Street, Central
Hong Kong, China
Tel: (852) 2530-1409
Fax: (852) 2526-1107
E-mail: sales@asia2000.com.hk
URL: http://www.asia2000.com.hk

HUNGARY
Euro Info Service
Margitszgeti Europa Haz
H-1138 Budapest
Tel: (36 1) 350 80 24, 350 80 25
Fax: (36 1) 350 90 32
E-mail: euroinfo@mail.matav.hu

INDIA
Allied Publishers Ltd.
751 Mount Road
Madras - 600 002
Tel: (91 44) 852-3938
Fax: (91 44) 852-0649

INDONESIA
Pt. Indira Limited
Jalan Borobudur 20
P.O. Box 181
Jakarta 10320
Tel: (62 21) 390-4290
Fax: (62 21) 390-4289

IRAN
Ketab Sara Co. Publishers
Khaled Eslamboli Ave., 6th Street
Delafrooz Alley No. 8
P.O. Box 15745-733
Tehran 15117
Tel: (98 21) 8717819; 8716104
Fax: (98 21) 8712479
E-mail: ketab-sara@neda.net.ir

Kowkab Publishers
P.O. Box 19575-511
Tehran
Tel: (98 21) 258-3723
Fax: (98 21) 258-3723

IRELAND
Government Supplies Agency
Oifig an tSoláthair
4-5 Harcourt Road
Dublin 2
Tel: (353 1) 661-3111
Fax: (353 1) 475-2670

ISRAEL
Yozmot Literature Ltd.
P.O. Box 56055
3 Yohanan Hasandlar Street
Tel Aviv 61560
Tel: (972 3) 5285-397
Fax: (972 3) 5285-397

R.O.Y. International
PO Box 13056
Tel Aviv 61130
Tel: (972 3) 649 9469
Fax: (972 3) 648 6039
E-mail: royil@netvision.net.il
URL: http://www.royint.co.il

Palestinian Authority/Middle East
Index Information Services
P.O.B. 19502 Jerusalem
Tel: (972 2) 6271219
Fax: (972 2) 6271634

ITALY, LIBERIA
Licosa Commissionaria Sansoni SPA
Via Duca Di Calabria, 1/1
Casella Postale 552
50125 Firenze
Tel: (39 55) 645-415
Fax: (39 55) 641-257
E-mail: licosa@ftbcc.it
URL: http://www.ftbcc.it/licosa

JAMAICA
Ian Randle Publishers Ltd.
206 Old Hope Road, Kingston 6
Tel: 876-927-2085
Fax: 876-977-0243
E-mail: irpl@colis.com

JAPAN
Eastern Book Service
3-13 Hongo 3-chome, Bunkyo-ku
Tokyo 113
Tel: (81 3) 3818-0861
Fax: (81 3) 3818-0864
E-mail: orders@svt-ebs.co.jp
URL:
http://www.bekkoame.or.jp/~svt-ebs

KENYA
Africa Book Service (E.A.) Ltd.
Quaran House, Mfangano Street
P.O. Box 45245
Nairobi
Tel: (254 2) 223 641
Fax: (254 2) 330 272

Legacy Books
Loita House
Mezzanine 1
P.O. Box 68077
Nairobi
Tel: (254) 2-330853, 221426
Fax: (254) 2-330854, 561654
E-mail: Legacy@form-net.com

KOREA, REPUBLIC OF
Dayang Books Trading Co.
International Division
783-20, Pangba Bon-Dong,
Socho-ku
Seoul
Tel: (82 2) 536-9555
Fax: (82 2) 536-0025
E-mail: seamap@chollian.net

Eulyoo Publishing Co., Ltd.
46-1, Susong-Dong
Jongro-Gu
Seoul
Tel: (82 2) 734-3515
Fax: (82 2) 732-9154

LEBANON
Librairie du Liban
P.O. Box 11-9232
Beirut
Tel: (961 9) 217 944
Fax: (961 9) 217 434
E-mail: hsayegh@librairie-du-liban.com.lb
URL: http://www.librairie-du-liban.com.lb

MALAYSIA
University of Malaya Cooperative
Bookshop, Limited
P.O. Box 1127
Jalan Pantai Baru
59700 Kuala Lumpur
Tel: (60 3) 756-5000
Fax: (60 3) 755-4424
E-mail: umkoop@tm.net.my

MEXICO
INFOTEC
Av. San Fernando No. 37
Col. Toriello Guerra
14050 Mexico, D.F.
Tel: (52 5) 624-2800
Fax: (52 5) 624-2822
E-mail: infotec@rtn.net.mx
URL: http://rtn.net.mx

Mundi-Prensa Mexico S.A. de C.V.
c/Rio Panuco, 141-Colonia
Cuauhtemoc
06500 Mexico, D.F.
Tel: (52 5) 533-5658
Fax: (52 5) 514-6799

NEPAL
Everest Media International Services
(P.) Ltd.
GPO Box 5443
Kathmandu
Tel: (977 1) 416 026
Fax: (977 1) 224 431

NETHERLANDS
De Lindeboom/Internationale
Publicaties b.v.-
P.O. Box 202, 7480 AE Haaksbergen
Tel: (31 53) 574-0004
Fax: (31 53) 572-9296
E-mail: lindeboo@worldonline.nl
URL: http://www.worldonline.nl/~lindeboo

NEW ZEALAND
EBSCO NZ Ltd.
Private Mail Bag 99914
New Market
Auckland
Tel: (64 9) 524-8119
Fax: (64 9) 524-8067

Oasis Official
P.O. Box 3627
Wellington
Tel: (64 4) 499 1551
Fax: (64 4) 499 1972
E-mail: oasis@actrix.gen.nz
URL: http://www.oasisbooks.co.nz/

NIGERIA
University Press Limited
Three Crowns Building Jericho
Private Mail Bag 5095
Ibadan
Tel: (234 22) 41-1356
Fax: (234 22) 41-2056

PAKISTAN
Mirza Book Agency
65, Shahrah-e-Quaid-e-Azam
Lahore 54000
Tel: (92 42) 735 3601
Fax: (92 42) 576 3714

Oxford University Press
5 Bangalore Town
Sharae Faisal
PO Box 13033
Karachi-75350
Tel: (92 21) 446307
Fax: (92 21) 4547640
E-mail: ouppak@TheOffice.net

Pak Book Corporation
Aziz Chambers 21, Queen's Road
Lahore
Tel: (92 42) 636 3222; 636 0885
Fax: (92 42) 636 2328
E-mail: pbc@brain.net.pk

PERU
Editorial Desarrollo SA
Apartado 3824, Ica 242 OF. 106
Lima 1
Tel: (51 14) 285380
Fax: (51 14) 286628

PHILIPPINES
International Booksource Center Inc.
1127-A Antipolo St, Barangay,
Venezuela
Makati City
Tel: (63 2) 896 6501; 6505; 6507
Fax: (63 2) 896 1741

POLAND
International Publishing Service
Ul. Piekna 31/37
00-677 Warzawa
Tel: (48 2) 628-6089
Fax: (48 2) 621-7255
E-mail: books%ips@ikp.atm.com.pl
URL:
http://www.ipscg.waw.pl/ips/export

PORTUGAL
Livraria Portugal
Apartado 2681, Rua Do Carm
o 70-74
1200 Lisbon
Tel: (1) 347-4982
Fax: (1) 347-0264

ROMANIA
Compani De Librarii Bucuresti S.A.
Str. Lipscani no. 26, sector 3
Bucharest
Tel: (40 1) 313 9645
Fax: (40 1) 312 4000

RUSSIAN FEDERATION
Isdatelstvo <Ves Mir>
9a, Kolpachniy Pereulok
Moscow 101831
Tel: (7 095) 917 87 49
Fax: (7 095) 917 92 59
ozimarin@glasnet.ru

**SINGAPORE; TAIWAN, CHINA
MYANMAR; BRUNEI**
Hemisphere Publication Services
41 Kallang Pudding Road #04-03
Golden Wheel Building
Singapore 349316
Tel: (65) 741-5166
Fax: (65) 742-9356
E-mail: ashgate@asianconnect.com

SLOVENIA
Gospodarski vestnik Publishing
Group
Dunajska cesta 5
1000 Ljubljana
Tel: (386 61) 133 83 47; 132 12 30
Fax: (386 61) 133 80 30
E-mail: repansekj@gvestnik.si

SOUTH AFRICA, BOTSWANA
For single titles:
Oxford University Press Southern
Africa
Vasco Boulevard, Goodwood
P.O. Box 12119, N1 City 7463
Cape Town
Tel: (27 21) 595 4400
Fax: (27 21) 595 4430
E-mail: oxford@oup.co.za

For subscription orders:
International Subscription Service
P.O. Box 41095
Craighall
Johannesburg 2024
Tel: (27 11) 880-1448
Fax: (27 11) 880-6248
E-mail: iss@is.co.za

SPAIN
Mundi-Prensa Libros, S.A.
Castello 37
28001 Madrid
Tel: (34 91) 4 363700
Fax: (34 91) 5 753998
E-mail: libreria@mundiprensa.es
URL: http://www.mundiprensa.com/

Mundi-Prensa Barcelona
Consell de Cent, 391
08009 Barcelona
Tel: (34 3) 488-3492
Fax: (34 3) 487-7659
E-mail: barcelona@mundiprensa.es

SRI LANKA, THE MALDIVES
Lake House Bookshop
100, Sir Chittampalam Gardiner
Mawatha
Colombo 2
Tel: (94 1) 32105
Fax: (94 1) 432104
E-mail: LHL@sri.lanka.net

SWEDEN
Wennergren-Williams AB
P. O. Box 1305
S-171 25 Solna
Tel: (46 8) 705-97-50
Fax: (46 8) 27-00-71
E-mail: mail@wwi.se

SWITZERLAND
Librairie Payot Service Institutionnel
C(tm)tes-de-Montbenon 30
1002 Lausanne
Tel: (41 21) 341-3229
Fax: (41 21) 341-3235

ADECO Van Diermen
EditionsTechniques
Ch. de Lacuez 41
CH1807 Blonay
Tel: (41 21) 943 2673
Fax: (41 21) 943 3605

THAILAND
Central Books Distribution
306 Silom Road
Bangkok 10500
Tel: (66 2) 2336930-9
Fax: (66 2) 237-8321

**TRINIDAD & TOBAGO
AND THE CARRIBBEAN**
Systematics Studies Ltd.
St. Augustine Shopping Center
Eastern Main Road, St. Augustine
Trinidad & Tobago, West Indies
Tel: (868) 645-8466
Fax: (868) 645-8467
E-mail: tobe@trinidad.net

UGANDA
Gustro Ltd.
PO Box 9997, Madhvani Building
Plot 16/4 Jinja Rd.
Kampala
Tel: (256 41) 251 467
Fax: (256 41) 251 468
E-mail: gus@swiftuganda.com

UNITED KINGDOM
Microinfo Ltd.
P.O. Box 3, Omega Park, Alton,
Hampshire GU34 2PG
England
Tel: (44 1420) 86848
Fax: (44 1420) 89889
E-mail: wbank@microinfo.co.uk
URL: http://www.microinfo.co.uk

The Stationery Office
51 Nine Elms Lane
London SW8 5DR
Tel: (44 171) 873-8400
Fax: (44 171) 873-8242
URL: http://www.the-stationery-office.co.uk/

VENEZUELA
Tecni-Ciencia Libros, S.A.
Centro Cuidad Comercial Tamanco
Nivel C2, Caracas
Tel: (58 2) 959 5547; 5035; 0016
Fax: (58 2) 959 5636

ZAMBIA
University Bookshop, University of
Zambia
Great East Road Campus
P.O. Box 32379
Lusaka
Tel: (260 1) 252 576
Fax: (260 1) 253 952

ZIMBABWE
Academic and Baobab Books (Pvt.)
Ltd.
4 Conald Road, Graniteside
P.O. Box 567
Harare
Tel: 263 4 755035
Fax: 263 4 781913

www.ingramcontent.com/pod-product-compliance
Lightning Source LLC
Chambersburg PA
CBHW080237270326
41926CB00020B/4269